MANUAL OF DISSECTION

MANUAL OF DISSECTION

Branislav Vidić, S.D.

Professor of Anatomy,
Georgetown University Schools
of
Medicine and Dentistry,
Washington, D.C.

with 17 illustrations

The C. V. Mosby Company

ST. LOUIS TORONTO PRINCETON 1984

MOSBY

A TRADITION OF PUBLISHING EXCELLENCE

Editor: Samuel E. Harshberger
Assistant editor: Anne Gunter
Manuscript editor: Marjorie L. Sanson
Book design: Kay M. Kramer
Production: Linda R. Stalnaker, Judy England, Margaret B. Bridenbaugh

Printed in the United States of America

The C.V. Mosby Company
11830 Westline Industrial Drive, St. Louis, Missouri 63146

Library of Congress Cataloging in Publication Data

Vidić, Branislav.
Manual of dissection.

Includes index.
1. Human dissection. 2. Anatomy, Human—Laboratory manuals. I. Title.
QM34.V5 1984 611'.0028 83-13259
ISBN 0-8016-5245-6

AC/VH/VH 9 8 7 6 5 4 3 2 1 01/A/083

Preface

The quality and the educational result of laboratory work in anatomy are greatly dependent on the preparedness and guidance of students to perform independently regional dissections. This practical experience is essential for the subsequent elaboration of whole body composition and function. Because dissection is so important in the anatomy curriculum, students should be supplied in the course of their laboratory exercise, in addition to the oral instruction, with visual teaching aids, such as written guidance for dissecting procedures, slides, films, photographs, and drawings.

The *Manual of Dissection* provides technical instruction for the stepwise development of regional dissections; the descriptive anatomical insertions are omitted throughout. The anatomical structures are identified in accordance with the International Nomenclature, *Nomina Anatomica,* ed. 4, 1977, or its English translation. The text is concise and composed in the logical sequence of dissection. The *Title* identifies the anatomical region and structures. *Observation* and *Palpation* provide instruction on which surface anatomical landmarks should be identified and how they may be used for the initiation and the course of subsequent dissection. The textural instruction for the initiation of dissection is always complemented by a drawing that diagramatically indicates the points or lines of a regional skin incision. *Dissection,* which usually follows a sequence from the surface inward, contains procedural information on how to expose, preserve, and remove the structures in the course of a regional study. *Special notes* are inserted wherever appropriate to emphasize the sexual differences, technical procedures, or the need to observe simultaneously the structures from more than one cadaver. For the convenience of the reader the muscles of each dissecting assignment are grouped in tables and their origins, insertions, innervations, and actions indicated. At the end of each regional dissection students are advised to consult the *Summary of structures,* which provides a complete list of bones, cartilages, central nervous system structures, fasciae, ganglia, glands, joints, ligaments, lymph nodes, membranes, muscles, nerves, organs, plexuses, surface anatomical structures, and vessels in that particular region.

Numerous page cross-references to the appropriate skeletal, surface anatomical, or dissecting views in the *Photographic Atlas of the Human Body* are included in the Manual to assist students in the following ways:

1. Preview the structures and relationships that subsequently will be either identified as the surface anatomical landmarks or dissected in the laboratory material
2. Control the dissecting sequence
3. Compare between their dissections and the corresponding views in the Atlas
4. Review the regional anatomy

Because the sequence of the Manual follows closely that of the Atlas, it is ideally suited for independent dissection in the anatomical laboratory.

All drawings were prepared by June L. Melloni, M.A., Medical Illustrator. Roy R. Peterson, Ph.D., Professor of Anatomy at Washington University School of Medicine reviewed and updated the anatomical terminology for the manual. Contributions of many colleagues, students, and members of The C.V. Mosby Company in the completion of the *Manual of Dissection* were essential.

Branislav Vidić

Contents

Part Six

PERINEUM

Part Seven

LOWER LIMB

MANUAL OF DISSECTION

Introduction

GENERAL INSTRUCTIONS

The practical work in anatomy is based on the organized dissection of regions of the human body. Each region usually consists of a restricted area of the skin, limited by either conventional or anatomical boundaries, and those tissue layers that are contained between the skin and the skeleton. Before a dissection is begun, it is advisable to preview the anatomical structures of a given region (consult the *Atlas*) and the technical directions about a dissection (consult the *Manual*). The laboratory work is best achieved through the following consecutive procedures.

Observation. Prior to a dissection, all noteworthy features should be observed on the surface of the body. These include the skeletal prominences, cartilages, muscles, tendons, subcutaneous veins, natural openings, and so on.

Palpation. Many anatomical elements that are not visible on the surface of the body may be identified by palpation (digital identification), particularly in a living person. These include bones, bony openings and prominences, joints, cartilages, muscles, tendons, and occasionally blood vessels, lymph nodes, and viscera. Some internal structures that are not observable or palpable may be identified through the natural openings (e.g., prostate, palatine tonsil).

Dissection. After the anatomical structures have been identified and their relationships determined through observation and palpation, the dissection may commence. For this procedure several instruments are needed: scalpel, forceps, sound, scissors, syringe, needles, probes, directories, saw, chisel, and hammer. Whatever the selected region, the dissection procedure should demonstrate the anatomical structures layer by layer. In the course of a dissection from the surface to the skeleton the elements should be preserved, except as otherwise indicated.

The first dissecting step is the reflection of the skin, i.e., the laying back or peeling away of an area of skin limited by the incisions. The epidermis and the dermis are cut along the indicated lines and gradually removed from the intact subcutaneous tissue. Once the latter tissue is exposed, the dissection of superficial elements, such as nerves and veins superficial to the fascia, can be done. In the face, neck, and hand, however, the subcutaneous muscles are also found. Following the dissection of subcutaneous elements the fascial surface should be exposed within the limits of a given region. At this time the openings in the fascia through which blood vessels and nerves pass from deep to superficial, or vice versa, may be observed. The fascia is subsequently cut along the lines of the cutaneous incisions and reflected to the same side as the skin. Then the structures deep to the fascia can be systematically evaluated as far as the skeleton.

After one region is completely dissected, the dissection of the next consecutive region is begun by the same methods. Each part of the body (head, neck, upper limb, thorax, abdomen, pelvis, perineum, and lower limb) is composed of a number of regions. Only when all the regions of a body part are properly demonstrated should the bony attachments of muscles, bones, and joints ultimately be studied.

SPECIFIC INSTRUCTIONS

Although this manual could be used independently for technical instructions relative to the dissection of body regions, it is integrated in content and sequence with the *Photographic Atlas of the Human Body*. Throughout the manual references are made to appropriate pages of the atlas to facilitate the appreciation of the anatomical region under consideration and the application of technical instruction from the manual to the cadaveric material. It is advisable, therefore, to preview the sequential illustrations of a region in the atlas before the dissection of that region is initiated. This approach provides students with sufficient information about the regional anatomical structures, the relationships between them, and their disposition in sequential layers from the surface toward the skeleton. Consequently students should be informed well enough beforehand to plan and proceed with a new dissection in such a way as to identify and demonstrate an optimum amount of anatomical structures. Drawings in the manual should prove helpful for initiation of a new dissection in guiding students to identify the regional boundaries and to plan the skin incision.

PREPARATION AND MAINTENANCE OF CADAVER

Preparation of cadavers for the laboratory exercise is a procedure through which all organic substrates of and pathogenic materials in cadavers become fixed with a mixture of chemical agents. The fixative consists of water (about 57.5%), alcohol (about 30%), glycerin (about 5%), phenol (about 5%), and formaldehyde (about 2.5%). After the femoral artery is exposed on one side of the body, the fixative (4 to 10 gallons total) is induced through this artery into the arterial system, heart, pulmonary blood vessels, and veins under a pressure of 10 to 12 pounds of water. Only those parts of the body which have not been sufficiently infiltrated by fixative are additionally treated by local vascular or subcutaneous injection of fixative. Perfused cadavers are subsequently maintained in plastic bags at 0° C for at least 15 weeks prior to the beginning of dissection.

The practical work in the gross anatomy laboratory extends through a number of months. It is necessary to maintain the cadaver in good condition throughout this period so that, at the end, the dissection may be as effective as at the beginning. Hence it is important that, at the end of each session, the dissected area be covered again by the reflected skin, which may be clipped to the margin of the intact skin. A few layers of wet paper toweling should be left over the skin incisions until the next session. All pieces of the cadaver that were taken out during the dissection should be preserved in the waste receptacle and the tables cleaned. Finally, the plastic bag should be closed over the cadaver.

Part One
HEAD AND NECK

ORBICULARIS OCULI MUSCLE

Observation. Observe the palpebral fissure, the medial and lateral angles of the eye, the folds that diverge from the lateral angle, the supraorbital ridge, the cilia, and the eyebrows (p. 23).

Palpation. Identify by this procedure the margin of the orbital opening (aditus orbitae), the supraorbital notch (foramen), and the infraorbital foramen (p. 2). Reflect carefully first the upper and then the lower eyelid with forceps, and observe the posterior or deep aspect of the eyelid covered by the conjunctiva.

Dissection. Cut the skin (epidermis and dermis) around the orbit about 2 cm from the bony margins (Fig. 1). Gradually reflect the skin, dissecting all around from the periphery toward the palpebral fissure. While reflecting the skin, dissect the circular muscular fibers of the orbicularis oculi muscle, one at a time and in the same direction. Peripheral circular fibers, which overlap the bony boundaries of the orbit, form the orbital part of the muscle. The subcutaneous tissue here is represented by a more or less developed fatty layer, which should be gradually removed from the surface

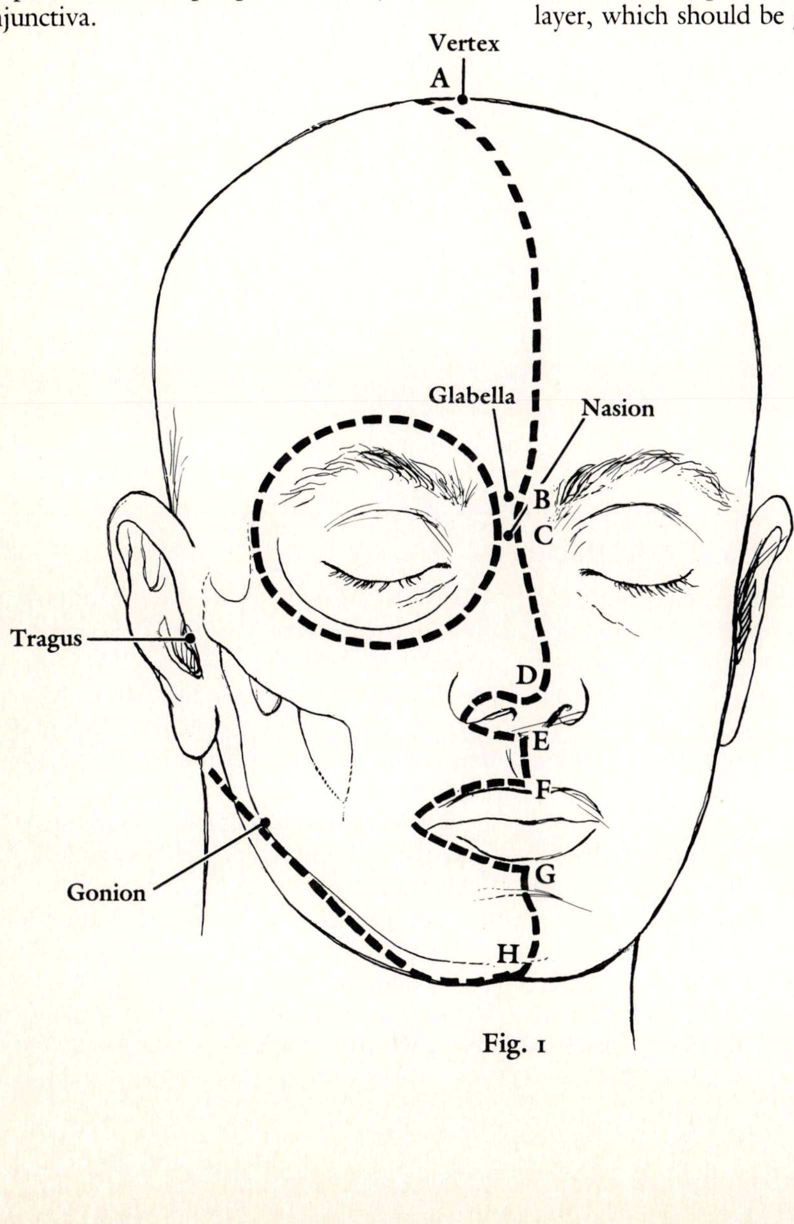

Fig. 1

Table 1

Name	Origin	Insertion	Innervation	Action
Orbicularis oculi	Nasal part of frontal bone, frontal process of maxilla, and medial palpebral ligament	Skin of eyebrow and lateral palpebral raphe	Temporal and zygomatic branches of facial nerve	Sphincter of eyelids

of the muscle. The central part of the muscle, which is composed of circular and semicircular fibers within both eyelids, is the palpebral part. Because there is no fatty tissue interposed between the dermis and this muscular part, care must be taken here when reflecting the skin not to damage the muscular fibers. Remove the skin already reflected toward the palpebral fissure by cutting along the free margins of the upper and lower eyelids, from one angle to the other. Study the surface of the muscle and the attachments of both muscular parts, and ascertain the possible effect of the muscular action (p. 24). Expose the medial palpebral ligament and, on the surface of the muscle close to the medial angle of the palpebral fissure, the angular artery and vein, a branch of the facial artery, and a tributary of the facial vein.

Summary of structures

1. Arteries
   Angular
   Facial
2. Bones and bony landmarks
   Infraorbital foramen
   Orbital opening
   Supraorbital notch (foramen)
3. Ligament
   Medial palpebral
4. Muscles
   Orbicularis oculi, orbital part
   Orbicularis oculi, palpebral part
5. Surface anatomy
   Cilia
   Conjunctiva
   Eyebrows
   Eyelids
   Lateral palpebral angle
   Medial palpebral angle
   Palpebral fissure
   Superciliary ridge
6. Veins
   Angular
   Facial

PAROTID, TEMPORAL, AND MASSETERIC REGIONS

Observation. Observe and locate precisely vertex, glabella, nasion, zygoma, chin, gonion, tip of the mastoid process, tragus, zygomatic arch, base of the mandible, posterior border of the mandibular ramus, nose, mouth, ear, and facial skinfolds (pp. 2, 3, and 27).

Palpation. Palpate around the nasal opening the septal, lateral and greater alar cartilages of the external nose (pp. 120 and 121).

Reflect first the upper and then the lower lip and observe the vestibule of the mouth, between the lips and cheeks on one side and the alveolar arches and teeth on the other.

Palpate the masseter and temporal muscles and the parotid gland, from the posterior border of the mandibular ramus backward to the mastoid process and upward to the auricle. On the lateral, or superficial, aspect of the masseter muscle attempt to palpate the parotid duct at the level of a line from the junction of the ala nasi with the face to a point about 1 cm above the gonion.

The movement of the mandibular head in the temporomandibular joint may be palpated during closing and opening of the mouth by introducing the index finger into the external acoustic meatus and pressing the finger forward.

Dissection. Incise the skin (Fig. 1) from the vertex down over the glabella, nasion, and dorsum of the nose as far as the tip of the nose. Proceed around the nasal opening to below the nasal septum, and follow the midline down to the red margin of the upper lip. Cut along this margin around to the midpoint of the lower lip, and then vertically down to the chin. From the chin continue along the base of the mandible, over the gonion, to a point 2 to 3 cm posterior to the gonion. Reflect the skin from the chin posteriorly to the level of a vertical line extending from the vertex, through the tragus, and terminating at a point again 2 to 3 cm posterior to the gonion.

Dissect the zygomaticus muscle along a line leading from the zygomatic bone to the corresponding angle of the mouth, and the risorius and depressor anguli oris muscles from the angle of the mouth posterior and posteroinferiorly to the base of the mandible (p. 28).

On the lateral aspects of the parotid gland and the masseter muscle expose the upper attachment of the platysma, which from there extends inferiorly into the neck. Reflect the dissected part of the platysma downward to the level of the skin incision. Gradually expose the surface of the parotid gland from posterior to anterior. In front of the tragus dissect the superficial temporal artery and vein and the auriculotemporal nerve. The pulsation of this artery may be felt in a living subject by slightly compressing the vessel against the skeletal background. Extend the dissection of the lateral surface of the parotid gland anteriorly. Notice its irregular anterior border, from which the initial part of the parotid duct should be exposed. Follow the duct anteriorly, over the lateral aspect of the masseter muscle, and observe how it curves around the anterior border of the latter muscle and then proceeds internally. Dissect the duct as deep as the surface of the buccinator muscle, through which the duct passes first to open into the oral vestibule opposite the maxillary second molar (p. 30).

Along the anterior border of the gland isolate numerous branches of the facial nerve and the transverse facial artery. This vessel passes between and parallel to the parotid duct and the zygomatic arch. From here the branches (temporal, zygomatic, buccal, marginal mandibular, and cervical, from superior inferiorly) extend into the face, where they should be followed to the posterior borders of the zygomaticus and depressor anguli oris muscles.

Dissect the lateral surface of the masseter muscle, from the zygomatic arch to the angle of the mandible (p. 30). Observe a superficial, oblique part and a deep, vertical part of this muscle. In front of the anterior border of the muscle, close to the mandibular attachment, dissect the facial artery and vein, which enter the face from the neck by passing around a notch on the base of the mandible. The pulsations of the facial artery may be felt in a living subject by compressing the vessel against the notch.

Cut the temporal fascia horizontally, parallel and close to the upper border of the zygomatic arch, and remove it from the muscle. Dissect the entire surface of the temporalis muscle in the temporal fossa down to the level of the zygomatic arch.

Table 2

Name	Origin	Insertion	Innervation	Action
Buccinator	Outer aspects of alveolar processes of maxilla and mandible opposite molars and pterygomandibular raphe	Angle of mouth and lips	Buccal branch of facial nerve	Compresses cheeks against teeth
Depressor anguli oris	Oblique line of mandible	Angle of mouth	Mandibular marginal branch of facial nerve	Depresses angle of mouth
Masseter	Zygomatic process of maxilla and zygomatic arch	Middle and lower halves and angle of the lateral aspect of mandibular ramus	Anterior trunk of mandibular nerve	Elevates, protracts, and retracts mandible
Platysma	Pectoralis major and deltoid fasciae	Symphysis menti, lower border of mandibular body, and skin of lower part of face	Cervical branch of facial nerve	Depresses mandible, lower lip, and angle of mouth; deep inspirator; diminishes concavity between mandible and side of neck
Risorius	Parotid fascia	Angle of mouth	Buccal branch of facial nerve	Retracts angle of mouth
Temporalis	Temporal fossa and deep aspect of temporal fascia	Coronoid process of mandible	Temporal branches of mandibular nerve	Elevates and retracts mandible
Zygomaticus	Zygomatic bone	Angle of mouth	Buccal branch of facial nerve	Elevates and retracts angle of mouth

Cut the parotid duct vertically, close to its origin, and reflect it anteriorly.

In front of the masseter muscle and on a deeper plane, expose the surface of the buccinator muscle as far anteriorly as the angle of the mouth (p. 32). Observe the relationship of the parotid duct and the ramifications of the buccal nerve, a branch of the mandibular nerve, on the surface of this muscle. Remove piecemeal the parotid tissue while following the blood vessels and nerves (superficial temporal artery and vein, transverse facial artery, auriculotemporal nerve, and branches of the facial nerve) from the surface into the gland. Find the temporofacial and cervicofacial trunks and, farther along, the junction of these trunks with the facial nerve. Expose the latter nerve superiorly (toward the stylomastoid foramen) and identify the two respective motor branches of it for the stylohyoid and the posterior belly of the digastric muscle. Dissect the auriculotemporal nerve through the parotid gland until it passes deep to the neck of the mandible. Notice the connections between the auriculotemporal and facial nerves.

Dissect the superficial temporal vein downward and find its junction with the maxillary vein, or the pterygoid plexus. From here the retromandibular vein originates. Deep to the latter vein, and still in the parotid region, expose the terminal part of the external carotid artery and its terminal branches, the maxillary and the superficial temporal arteries. From the latter artery the origin of the transverse facial branch may now be identified.

Summary of structures

1. Arteries
   External carotid
   Facial
   Maxillary
   Superficial temporal
   Transverse facial
2. Bones and bony landmarks
   Alveolar arches
   Base of mandible
   Glabella
   Gonion
   Mandibular angle
   Mandibular head
   Mandibular neck
   Mandibular ramus
   Mastoid process
   Nasion
   Stylomastoid foramen
   Teeth
   Temporal fossa
   Vertex
   Zygoma
   Zygomatic arch
3. Cartilages
   Greater alar
   Septal
   Triangular
4. Fasciae
   Parotidomasseteric
   Temporal
5. Glands
   Parotid
   Parotid duct
6. Joint
   Temporomandibular
7. Muscles
   Buccinator
   Depressor anguli oris
   Masseter
   Platysma
   Risorius
   Temporal
   Zygomatic
8. Nerves
   Auriculotemporal
   Buccal branch of facial
   Buccal branch of mandibular
   Cervical branch of facial
   Cervicofacial trunk
   Facial
   Marginal mandibular branch of facial
   Nerve to posterior belly of digastric muscle of facial
   Nerve to stylohyoid muscle of facial
   Temporal branch of facial
   Temporofacial trunk
   Zygomatic branch of facial
9. Surface anatomy
   Ala nasi
   Auricle
   Cheek
   Chin
   Ear
   External acoustic meatus
   Facial skinfolds
   Lips
   Mouth
   Nasal septum
   Nose
   Oral angle
   Oral vestibule
   Tragus
10. Veins
   Facial
   Maxillary
   Pterygoid plexus
   Retromandibular
   Superficial temporal

FACE AND ANTERIOR HALF OF SCALP

Dissection. While dissecting the muscles of the face (p. 34), complete the dissection of the branches of the facial nerve and the facial artery and vein all the way through the face, and follow the vessels to the medial angle of the palpebral fissure. Isolate the collateral branches of the facial artery for the upper and lower lips—the superior and inferior labial arteries.

In the region of the forehead dissect the vertical muscular fibers of the frontalis muscle and follow them posteriorly to their continuation, the galea aponeurotica.

Immediately medial to the zygomaticus muscle and on the same plane, dissect the levator labii superioris muscle from the inferior orbital margin, where this muscle is covered by the inferior part of the orbicularis oculi muscle, to the upper lip.

Immediately medial to the depressor anguli oris muscle and on the same plane, expose the depressor labii inferioris muscle from the lower lip down to the mandibular base. Between the right and left depressor labii inferioris muscles, observe the fibers of the mentalis muscle, which are directed from the mandible to the deep aspect of the skin.

Reflect the parts of the orbicularis oculi and frontalis muscles from the medial half of the superior orbial margin and identify the supraorbital artery, vein, and nerve in the corresponding foramen (p. 36). Trace these structures through the forehead. Just medial to the supraorbital neurovascular bundle notice the supratrochlear artery, vein, and nerve, which curve around the medial part of the superior orbital margin on their way from the orbit into the face.

Remove the levator labii superioris muscle from the inferior orbital margin and reflect it down to the upper lip. Deep to this muscle find the infraorbital foramen and its contents—the infraorbital artery, vein, and nerve. Dissect these components individually. Deep to the infraorbital vessels and nerve expose the levator anguli oris muscle (p. 36) from its bony attachment down to the corresponding angle of the mouth. Observe how the labial muscles, the levators and the depressors, interdigitate with the circular fibers of the orbicularis oris muscle.

Remove the depressor anguli oris and depressor labii inferioris muscles, starting from the posterior border of the former, and reflect them anteriorly until the mental foramen and its contents, the mental artery, vein, and nerve, are exposed (p. 36).

Reflect first the upper and then the lower lip and remove the mucous membrane carefully from their posterior, ves-

Table 3

Name	Origin	Insertion	Innervation	Action
Depressor anguli oris (see Table 2)				
Depressor labii inferioris	Oblique line of mandible	Skin of lower lip	Mandibular marginal branch of facial nerve	Depresses lower lip
Frontalis	Subcutaneous layer of root of nose, supraorbital margin, and zygomatic process of frontal bone	Galea aponeurotica	Temporal branch of facial nerve	Elevates eyebrow and skin of root of nose
Levator anguli oris	Canine fossa	Angle of mouth	Buccal branch of facial nerve	Elevates angle of mouth
Levator labii superioris	Infraorbital margin	Upper lip	Buccal branch of facial nerve	Elevates upper lip
Mentalis	Incisive fossa	Skin of chin	Mandibular marginal branch of facial nerve	Elevates and protrudes lower lip; wrinkles skin of chin
Orbicularis oculi (see Table 1)				
Orbicularis oris	Skin of lips	Mucous membrane of lips; maxilla and mandible	Buccal and mandibular marginal branches of facial nerve	Sphincter of lips; compresses lips against teeth; protrudes lips; articulation and mastication
Zygomaticus (see Table 2)				

tibular aspect. Observe the labial salivary glands just deep to the labial mucous membrane.

While dissecting the branches of the facial nerve in the face, note the many connections between these branches and those of the trigeminal nerve: the supraorbital, the infraorbital, and the mental.

Summary of structures

1. Arteries
   Facial
   Inferior labial
   Infraorbital
   Mental
   Superior labial
   Supraorbital
   Supratrochlear
2. Bones and bony landmarks
   Base of mandible
   Infraorbital foramen
   Infraorbital margin
   Mandible
   Mental foramen
   Orbit
   Supraorbital foramen
   Supraorbital margin
3. Fascia
   Galea aponeurotica
4. Gland
   Labial salivary
5. Muscles
   Depressor anguli oris
   Depressor labii inferioris
   Frontalis
   Levator anguli oris
   Levator labii superioris
   Mentalis
   Orbicularis oculi
   Orbicularis oris
   Zygomaticus
6. Nerves
   Branches of facial
   Facial
   Infraorbital
   Mental
   Supraorbital
   Supratrochlear
   Trigeminal
7. Surface anatomy
   Face
   Lower lip
   Medial palpebral fissure
   Oral angle
   Upper lip
8. Veins
   Facial
   Infraorbital
   Mental
   Supraorbital
   Supratrochlear

INFRATEMPORAL FOSSA AND TEMPOROMANDIBULAR JOINT

Observation. Observe the bony walls of the infratemporal fossa on the skelton (pp. 3 and 4).

Palpation. Locate by palpation the exact position of the bony components of the temporomandibular joint and the coronoid process of the mandible, with the mouth open and closed.

Dissection. Cut vertically, along the posterior border of the mandibular ramus, the branches of the facial nerve and transverse facial artery, and reflect them anteriorly. Use the chisel or saw to cut the zygomatic arch, anteriorly just behind the body of the zygomatic bone and posteriorly just in front of the articular tubercle of the temporal bone (p. 3). Remove the masseter muscle together with the zygomatic arch (p. 38) gradually, toward the mandible, and expose the masseteric artery, vein, and nerve, which penetrate the deep muscular aspect from the infratemporal fossa by way of the mandibular notch. The masseteric vessels and nerve should be severed close to their point of entry into the muscle so that the muscle can be further reflected inferiorly.

Complete the dissection of the temporalis muscle down to the attachment to the coronoid process (p. 38). Simultaneously separate the buccal nerve, a branch of the mandibular nerve, from the tendon of this muscle as far superiorly as possible. Use a chisel or saw to cut obliquely the coronoid process from the bottom of the mandibular notch to the point just behind the mandibular third molar. Reflect the bony process, with the temporalis muscle attached to it, superiorly. Once in the infratemporal fossa (p. 40) dissect the maxillary artery, if the artery has a superficial position to the lateral pterygoid muscle. If it is located deep to this muscle, it should be dissected only after the muscle has been removed. Expose simultaneously the collateral branches of the artery—the middle meningeal, inferior alveolar, muscular, and superior posterior alveolar.

Expose the lateral surface of the upper head of the lateral pterygoid muscle (p. 40). Identify the middle and posterior deep temporal nerves, branches of the mandibular nerve, as they emerge from above this head and ascend onto the deep aspect of the temporalis muscle. At the same level complete the dissection of the masseteric nerve, also a branch of the mandibular nerve.

Expose the lateral surface of the lower head of the lateral pterygoid muscle (p. 40). In the space between the two heads notice two additional branches of the mandibular nerve. The anterior deep temporal branch ascends on the deep surface of the temporalis muscle, while the buccal nerve proceeds to the lateral surface of the bucinnator muscle.

Immediately inferior to the lower head, parallel and posterior to the buccal nerve, expose in succession the lingual and the inferior alveolar nerves. The latter nerve, before it enters the mandibular canal, gives off the mylohyoid nerve.

Remove completely both heads of the lateral pterygoid muscle (p. 42). Follow all the nerves that have already been identified toward the trunk of the mandibular nerve. Find the junction of the lingual nerve with the chorda tympani. Dissect the mandibular trunk up toward the foramen ovale and localize the position of the otic ganglion, which is situated immediately medial to and is more or less fused with the nerve trunk. Complete the dissection of the auriculotemporal nerve from the ganglion posteriorly until it passes medial to the mandibular neck.

Expose the lateral surface of the medial pterygoid muscle (p. 42).

By chiseling through the external surface of the mandible, it is possible to open the mandibular canal from the mandibular to the mental foramen. If this dissection is done, demonstrate, in the canal, the inferior alveolar vessels and nerve and their branches to the mandibular teeth.

The lateral surface of the articular capsule of the temporomandibular joint and the lateral temporomandibular ligament may now be demonstrated. Two additional ligaments of this joint extend from the base of the skull to the mandible: the sphenomandibular to the lingula of the mandible and the stylomandibular to the posterior border of the mandibular ramus. Both ligaments may be demonstrated by a dissection that begins behind and medial to the posterior border of the mandibular ramus. Remove the lateral half of the articular capsule from the temporal bone, mandible, and the intraarticular disc (p. 42). Demonstrate the articular surfaces of the joint and the disc between them. Notice the two compartments, upper and lower, of the joint. Study the movements of the mandibular head and the intraarticular disc while depressing, elevating, protracting, retracting, and circumducting the mandible.

Summary of structures

1. Arteries
   - Dental branches of inferior alveolar
   - Inferior alveolar
   - Masseteric
   - Maxillary
   - Middle meningeal
   - Muscular branches of maxillary
   - Superior posterior alveolar
   - Transverse facial
2. Bones and bony landmarks
   - Articular tubercle of temporal
   - Coronoid process of mandible
   - Foramen ovale
   - Infratemporal fossa
   - Lingula of mandible
   - Mandible
   - Mandibular canal
   - Mandibular head
   - Mandibular neck
   - Mandibular notch
   - Mandibular ramus
   - Mental foramen
   - Zygomatic

Table 4

Name	Origin	Insertion	Innervation	Action
Buccinator (see Table 2)				
Lateral pterygoid	Infratemporal crest of greater wing and lateral surface of lateral pterygoid plate	Neck of mandible and articular capsule of temporomandibular joint	Anterior trunk of mandibular nerve	Opens mouth; protrudes mandible
Masseter (see Table 2)				
Medial pterygoid	Medial surface of lateral pterygoid plate, pyramidal process of palatine bone, and tuberosity of maxilla	Medial aspect of ramus and angle of mandible	Mandibular nerve	Elevates and protrudes mandible
Temporalis (see Table 2)				

Zygomatic arch
3. Ganglion
Otic
4. Joints
Articular capsule
Articular surfaces
Intraarticular disk
Lateral temporomandibular ligament
Lower compartment
Sphenomandibular ligament
Stylomandibular ligament
Temporomandibular
Upper compartment
5. Muscles
Buccinator
Lateral pterygoid, lower head
Lateral pyerygoid, upper head
Masseter
Medial pterygoid
Temporalis
6. Nerves
Anterior deep temporal
Auriculotemporal
Branches of facial
Buccal nerve of mandibular
Chorda tympani
Dental branches of inferior alveolar
Inferior alveolar
Lingual
Mandibular
Masseteric
Middle deep temporal
Mylohyoid
Posterior deep temporal
7. Veins
Dental tributaries to inferior alveolar
Inferior alveolar
Masseteric

ANTERIOR AND POSTERIOR TRIANGLES OF THE NECK

Observation. Observe the thyroid cartilage, jugular notch, sternoclavicular joint, clavicle, acromioclavicular joint, sternocleidomastoid muscle, and possibly the external and anterior jugular veins (p. 45).

Palpation. Palpate the upper and lower borders of the thyroid cartilage, the thyrohyoid membrane above, the cricothyroid membrane below, the arch of the cricoid cartilage, the hyoid bone, the tracheal cartilages, the lobes of the thyroid gland on each side of the thyroid cartilage, the sternocleidomastoid muscle, and the anterior border of the trapezius muscle (p. 45).

Dissection. Incise the skin (Fig. 2) along the anterior and inferior limits of the triangles, i.e., from the midpoint of the chin vertically to the jugular notch, and then horizontally to the sternoclavicular joint, along the clavicle to the acromioclavicular joint. Reflect the skin posteriorly until the anterior free border of the trapezius muscle is evident. While reflecting the skin, dissect the platysma muscle (p. 46) in the subcutaneous tissue. Cut the muscle horizontally at the middle of the neck, using care not to injure any underlying structure, and reflect the halves upward and downward. In relation to the deep aspect of the upper muscular flap, complete the dissection of the cervical and the marginal mandibular branches of the facial nerve.

Immediately deep to the reflected platysma, starting from the parotid region, dissect the external jugular vein downward through the subcutaneous tissue until it pierces the fascia. In the anterior triangle between the two sternocleidomastoid muscles dissect, if evident, the anterior jugular vein downward, from the submental area through the subcutaneous tissue, until it pierces the fascia. Note the variable distribution and size of these veins by comparing them in several bodies.

Find the approximate midpoint of the posterior border of the sternocleidomastoid muscle. Cut the fascia at this level and spread it slightly upward and downward for about 1 cm. Deep to the fascia in this area find the terminal branches of the cervical plexus: transverse cervical, great auricular, lesser occipital, and supraclavicular nerves. These nerves may be traced from here as they pierce the fascia and extend farther in to the subcutaneous layer (p. 48).

From approximately the same midpoint at the sternocleidomastoid muscle, the external branch of the accessory nerve emerges from underneath this muscle to extend through the posterior triangle of the neck inferoposteriorly until it passes deep to the anterior border of the trapezius muscle. Note the lymph nodes along this course of this nerve.

Cut the fascia all along the sternocleidomastoid muscle, remove it, and expose the lateral aspect of the muscle. Cut the muscle horizontally through the middle and reflect the lower part downward. While reflecting the upper part superiorly, simultaneously follow the external branch of the accessory nerve under cover of or through the muscle. Find first the branch of this nerve to the sternocleidomastoid muscle and then the sternocleidomastoid artery. Both structures penetrate into the deep aspect of the muscle (p. 50).

Remove from the base of the mandible the fascia of the anterior triangle and subsequently expose the surface of the submandibular gland (superficial lobe). Note the close relationship between the upper border of the gland, the base of the mandible, and the facial artery and vein.

Deep to the sternocleidomastoid muscle, on the mastoid process, expose the attachment of the posterior belly of the digastric muscle and dissect it inferoanteriorly until it becomes continuous with the intermediate tendon, just above the greater horn of the hyoid bone. Dissect the stylohyoid muscle in front of and above the posterior belly as it extends from the styloid process to the hyoid bone. Note how the intermediate tendon of the digastric muscle becomes attached to the hyoid bone by the two bundles of the bifurcated stylohyoid muscle. The jugulodigastric lymph node may now be observed in proximity to the posterior border of the posterior digastric belly. Now complete the dissection of the branches of the facial nerve to the posterior belly of the digastric muscle and to the stylohyoid muscle. Expose also the anterior belly of the digastric muscle, following it from the intermediate tendon to the posterior aspect of the symphysis menti, or the digastric fossa, to which it attaches.

Expose successively on each side from the hyoid bone downward the anterior surfaces of the sternohyoid muscle, as far as the sternum, and of the three parts of the omohyoid muscle—superior belly, intermediate tendon, and inferior belly—until the latter belly becomes completely covered by the trapezius muscle.

Fig. 2

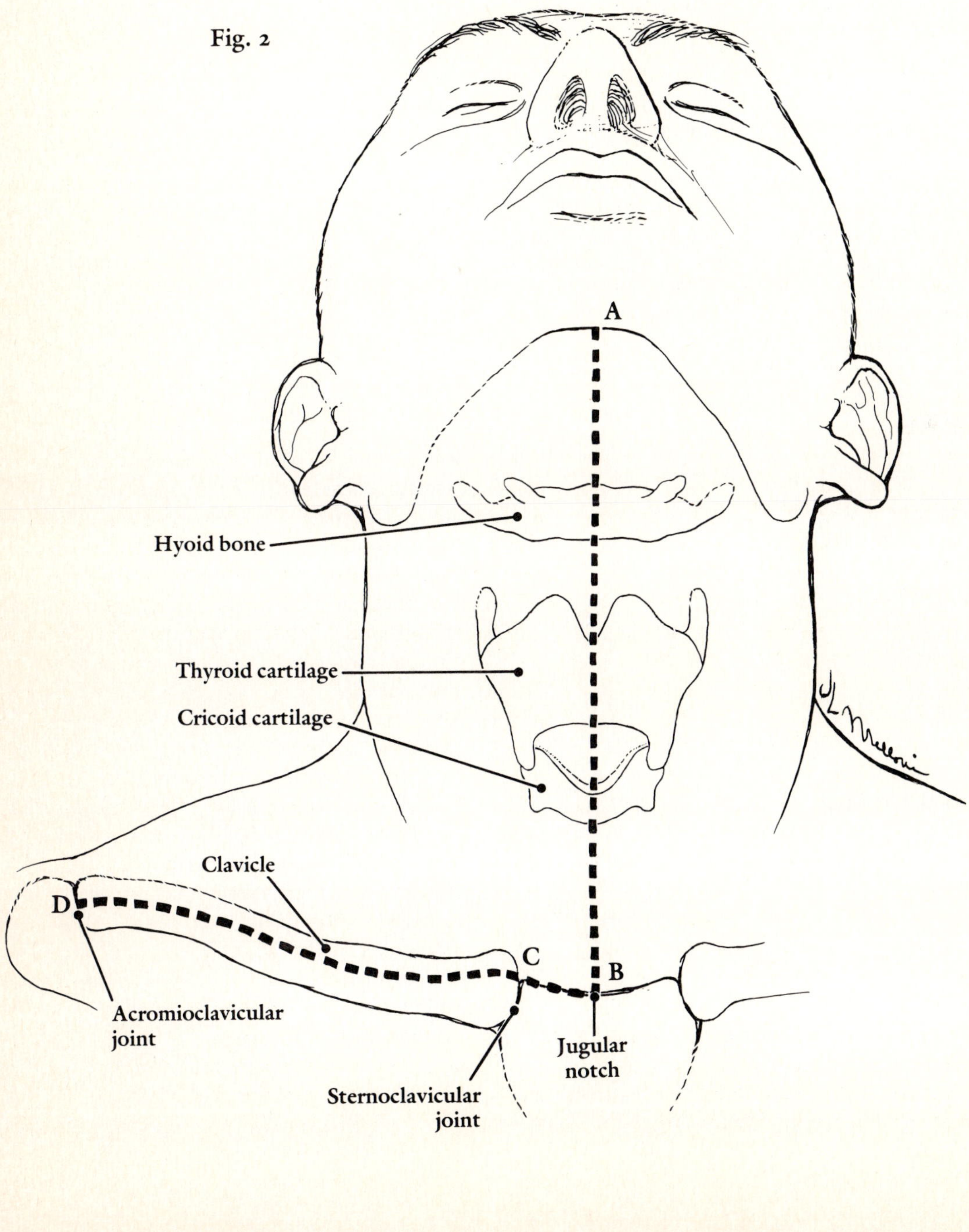

Table 5

Name	Origin	Insertion	Innervation	Action
Digastric	Mastoid notch, temporal bone	Digastric fossa of mandible	Mylohyoid branch of inferior alveolar nerve (anterior belly) and facial nerve (posterior belly)	Depresses mandible and elevates hyoid bone
Omohyoid	Upper border of scapula	Hyoid body	Superior ramus of ansa cervicalis	Depresses hyoid bone; active in inspiration
Platysma (see Table 2)				
Sternocleidomastoid	Manubrium sterni and medial third of clavicle	Mastoid process and superior nuchal line	Accessory nerve and ventral rami of C2 and C3 nerves	Laterally flexes head and rotates face to opposite side of action
Sternohyoid	Posterior surfaces of medial end of clavicle, sternoclavicular ligament, and manubrium sterni	Hyoid body	Ansa cervicalis	Depresses hyoid bone
Stylohyoid	Styloid process	Hyoid body	Facial nerve	Elevates and retracts hyoid bone
Trapezius	Medial third of superior nuchal line, external occipital protuberance, ligamentum nuchae, spinous processes, and supraspinous ligaments of cervical and thoracic vertebrae	Posterior border of lateral third of clavicle, medial margin of acromion, and superior lip of spine of scapula	Accessory nerve and ventral rami of C3 and C4 nerves	Stabilizes shoulder; elevates and rotates scapula; extends and laterally rotates head

Summary of structures

1. Arteries
   Facial
   Sternocleidomastoid
2. Bones and bony landmarks
   Clavicle
   Diagastric fossa
   Hyoid
   Hyoid, greater horn
   Jugular notch
   Mandibular base
   Mastoid process
   Sternum
   Styloid process of temporal
   Symphysis menti
3. Cartilages
   Cricoid arch
   Thyroid
   Tracheal
4. Fascia
   Cervical
5. Glands
   Submandibular
   Thyroid
6. Joints
   Acromioclavicular
   Sternoclavicular
7. Lymph nodes
   Jugulodigastric
   Superficial cervical
8. Membranes
   Cricothyroid
   Thyrohyoid
9. Muscles
   Digastric, anterior belly
   Digastric, intermediate tendon
   Digastric, posterior belly
   Omohyoid, inferior belly
   Omohyoid, intermediate tendon
   Omohyoid, superior belly
   Platysma
   Sternocleidomastoid
   Sternohyoid
   Stylohyoid
   Trapezius

10. Nerves
    - Accessory
    - Cervical branch of facial
    - Cervical plexus
    - Great auricular
    - Lesser occipital
    - Marginal mandibular branch of facial
    - Nerve to posterior belly of digastric muscle of facial
    - Nerve to stylohyoid muscle of facial
    - Supraclavicular
    - Transverse cervical
11. Surface anatomy
    - Anterior triangle of neck
    - Chin
    - Clavicle
    - Jugular notch
    - Posterior triangle of neck
    - Sternocleidomastoid muscle
    - Thyroid prominence
    - Trapezius muscle
12. Veins
    - Anterior jugular
    - External jugular
    - Facial

ANTERIOR AND POSTERIOR TRIANGLES OF THE NECK (continued)

Dissection. Identify the level at which the omohyoid muscle crosses the lateral aspect of the internal jugular vein. On the lateral aspect of the vein isolate the ansa cervicalis (p. 50); it is under cover of the muscle or sometimes above or below it. From this loop of nerves dissect superiorly its upper root and superoposteriorly its lower root. The latter passes across the lateral aspect of the vein and loops to the posterior surface of the vein. Observe the branches of the ansa cervicalis to the infrahyoid muscles.

Remove the anterior belly of the digastric muscle from the digastric fossa and the stylohyoid muscle from the hyoid bone. Reflect both muscles posteriorly and superiorly to the mastoid and styloid processes, respectively.

Dissect the internal jugular vein and its tributaries—the external and anterior jugular, superior thyroid, lingual, and facial veins in the neck.

Dissect the vagus nerve throughout the neck, posterior and internal to the internal jugular vein.

Immediately deep to the internal jugular vein (p. 52), from the level of the sternoclavicular joint superiorly, dissect the common carotid artery and identify its bifurcation. From here dissect further superiorly the external and internal carotid arteries. While exposing the external carotid artery in the neck, bring into evidence the initial parts of the collateral branches, from inferior to superior: superior thyroid artery, lingual artery, facial artery, and occipital artery. The latter gives off the previously observed sternocleidomastoid artery.

Reflect the submandibular gland superiorly onto the base of the mandible.

Dissect the surface of the mylohyoid muscle and observe on it the branches of the mylohyoid nerve and the submental artery, which courses parallel to the mandibular base. This is a branch of the facial artery.

From the junction of the upper root of the ansa cervicalis and the hypoglossal nerve follow the latter nerve anteriorly until it passes deep to the posterior free border of the mylohyoid muscle. Remove both mylohyoid muscles from the hyoid bone and reflect them upward (p. 52).

Complete the dissection of the deep lobe of the submandibular gland (that part located deep to the mylohyoid muscle) and observe the initial part of the submandibular duct from the glandular anterior pole.

Dissect the styloglossus muscle from the styloid process, which now may be easily palpated, to the side of the tongue. Along the anterosuperior border of the muscle find the lingual nerve and follow it anteriorly. At the point where the nerve is closely related to the submandibular gland, observe the submandibular ganglion. Note also how the nerve passes successively superior, lateral, inferior, and medial to the submandibular duct.

Complete the dissection of the hypoglossal nerve by following it toward the point of entry into the musculature of the tongue. Isolate its first branch, which extends into the thyrohyoid muscle. Many connections should be verified between the hypoglossal and lingual nerves. Finish the dissection of the submandibular duct until it pierces the mucous membrane of the mouth, and observe the relationship of the anterior segment of the duct with the sublingual gland.

Preserve the hypoglossal and lingual nerves and the submandibular duct while dissecting the hyoglossus muscle from the hyoid bone superiorly into the tongue, and the geniohyoid muscle from the anterior part of the hyoid bone to the mental, or genial, spine. Separate the hyoglossus and geniohyoid muscles from one another slightly and notice in between them the genioglossus muscle, which extends also from the mental spine into the tongue.

Follow the lingual artery from its origin until it passes deep to the posterior border of the hyoglossus muscle. Subsequently remove this muscle from the hyoid bone and finish the dissection of the underlying lingual artery.

Cut the sternohyoid muscle horizontally through the middle and reflect the halves superiorly and inferiorly. Dissect, under cover of this muscle, the sternothyroid muscle from the posterior aspect of the sternum to the oblique line of the thyroid cartilage, and the thyrohyoid muscle from the same oblique line to the hyoid bone superiorly (p. 50).

Continue the dissection of the superior thyroid artery (p. 52). Observe its first collateral branch, the superior laryngeal artery, which is accompanied by the internal branch of the superior laryngeal nerve. Follow these two structures until they pierce the thyrohyoid membrane to enter into the laryngeal interior. Follow now the internal branch of the superior laryngeal nerve superiorly to its junction with the external branch, and from there follow the superior laryngeal nerve, medial to the internal carotid artery, as far as the vagus nerve. Dissect the external branch from the bifurcation of the superior laryngeal nerve to where it ends in the cricothyroid muscle. In between both cricothyroid muscles the cricothyroid membrane, or ligament, may be demonstrated as it extends from the thyroid to the cricoid cartilage. Remove the thyrohyoid and omohyoid muscles from the hyoid bone and the sternothyroid muscle from the oblique line of the thyroid cartilage.

Expose and then study the lateral and anterior surfaces of the thyroid gland in situ.

Reflect the upper part of the internal carotid artery from the vertebral column at about the level of the gonion and cut the cervical fascia posterior to the artery. Expose the superior cervical sympathetic ganglion from underneath the fascia and follow the sympathetic trunk from the ganglion inferiorly. While preserving all arteries to which the trunk is closely related, namely the inferior thyroid and the vertebral, expose the middle ganglion along its course and, if accessible, the cervicothoracic, or stellate, ganglion.

Table 6

Name	Origin	Insertion	Innervation	Action
Cricothyroid	Anterior and lateral parts of cricoid cartilage	Inferior border and horn of thyroid cartilage	External branch of superior laryngeal nerve	Tenses and elongates vocal ligament
Digastric (see Table 5)				
Genioglossus	Upper genial tubercle	Lower, posterior, upper, and inferior parts of tongue	Hypoglossal nerve	Protrudes apex of tongue
Geniohyoid	Lower genial tubercle	Hyoid body	Ventral ramus of C1 nerve	Elevates hyoid bone; depresses mandible
Hyoglossus	Hyoid body and greater horn	Lateral part of tongue	Hypoglossal nerve	Depresses tongue
Mylohyoid	Mylohyoid line of mandible	Hyoid body and median raphe	Mylohyoid branch of inferior alveolar nerve	Elevates hyoid bone and oral floor; depresses mandible
Omohyoid (see Table 5)				
Sternohyoid (see Table 5)				
Sternothyroid	Posterior surface of manubrium sterni and cartilage of first rib	Oblique line of thyroid cartilage	Ansa cervicalis	Depresses larynx
Styloglossus	Styloid process	Lateral part of tongue	Hypoglossal	Retracts and elevates tongue
Stylohyoid (see Table 5)				
Thyrohyoid	Oblique line of thyroid cartilage	Hyoid body and greater horn	Ventral ramus of C1 nerve	Depresses hyoid bone; elevates larynx

Summary of structures

1. Arteries
   - Common carotid
   - External carotid
   - Facial
   - Inferior thyroid
   - Internal carotid
   - Lingual
   - Occipital
   - Sternocleidomastoid
   - Submental
   - Superior laryngeal
   - Superior thyroid
   - Vertebral
2. Bones and bony landmarks
   - Digastric fossa
   - Genial (mental) spine
   - Gonion
   - Hyoid
   - Mandibular base
   - Mastoid process
   - Styloid process of temporal
   - Vertebral column
3. Cartilages
   - Cricoid
   - Oblique line of thyroid
   - Thyroid
4. Fascia
   - Deep cervical
5. Ganglia
   - Middle cervical sympathetic
   - Stellate (cervicothoracic)
   - Submandibular
   - Superior cervical sympathetic
6. Glands
   - Sublingual
   - Submandibular
   - Submandibular duct
   - Thyroid
7. Joint
   - Sternoclavicular
8. Membranes
   - Cricothyroid
   - Thyrohyoid
9. Muscles
   - Cricothyroid
   - Digastric, anterior belly
   - Genioglossus
   - Geniohyoid
   - Hyoglossus
   - Infrahyoid
   - Mylohyoid
   - Omohyoid
   - Sternohyoid
   - Sternothyroid
   - Styloglossus
   - Stylohyoid
   - Thyrohyoid
10. Nerves
    - Ansa cervicalis
    - Ansa cervicalis, branches
    - Hypoglossal
    - Lingual
    - Lower root of ansa cervicalis
    - Mylohyoid
    - Nerve to thyrohyoid muscle
    - Superior laryngeal
    - Superior laryngeal, external branch
    - Superior laryngeal, internal branch
    - Sympathetic trunk
    - Upper root of ansa cervicalis
    - Vagus
11. Organs
    - Larynx
    - Tongue
12. Veins
    - Anterior jugular
    - External jugular
    - Facial
    - Internal jugular
    - Lingual
    - Superior thyroid

ANTERIOR AND POSTERIOR TRIANGLES OF THE NECK (continued)

Dissection. Remove completely the sternohyoid muscle and the inferior part of the sternocleidomastoid muscle.

Behind the sternoclavicular joint expose the junction between the internal jugular and subclavian veins. Dissect, on the left side of the body, the terminal portion of the thoracic duct. It ascends first behind the subclavian vein and, after curving over that vessel, opens into the subclavian vein, or the internal jugular vein, or the junction of these two veins.

Reflect back to their origins all branches of the cervical plexus (transverse cervical, great auricular, lesser occipital, supraclavicular) from those areas where they terminate, and continue exposing them toward the spinal cord. In the course of this procedure the ventral rami of the C1, C2, C3, and C4 spinal nerves, which form the plexus, should be identified.

On the anterior aspect of the scalenus anterior muscle (p. 54), which now may be observed through the fascia, expose the phrenic nerve, the ventral ramus of the C4 spinal nerve. Parallel and medial to this nerve the ascending cervical artery should first be identified and then dissected inferiorly as far as its origin from the inferior thyroid artery, or from the thyrocervical trunk. Continue the dissection along the thyrocervical trunk to the subclavian artery and isolate the following branches of the trunk: the inferior thyroid artery coursing to the thyroid gland (notice the relationship of this artery with the sympathetic trunk), the ascending cervical artery, the transverse cervical artery coursing horizontally backward, and the suprascapular artery coursing parallel and inferior to the transverse cervical artery.

Observe on the inferior aspect of the subclavian artery the beginning of the internal thoracic artery.

Deep to the arch of the inferior thyroid artery, between the scalenus anterior and the longus colli muscles, expose the vertebral artery and dissect it inferiorly to its origin from the subclavian artery. The vertebral artery is accompanied here by a branch of the sympathetic trunk, the vertebral nerve.

Complete the dissection of the subclavian vein while removing all of its tributaries.

Continue to dissect the subclavian artery behind the scalenus anterior muscle and expose the following collateral branches: the costocervical trunk, which divides into the deep cervical and highest intercostal arteries, and the dorsal scapular artery. Trace the subclavian artery in between the scalenus anterior and medius muscles.

Dissect the anterior surface of the scalenus anterior muscle.

In the intermuscular space between the scalenus anterior and medius muscles, in addition to the subclavian artery, dissect the trunks of the brachial plexus.

Dissect the anterior surface of the scalenus medius and posterior muscles, which are to some degree fused together. Notice several collateral branches of the brachial plexus as they are passing through the scalenus medius and posterior muscular mass. These are, from superior to inferior, nerve to the rhomboideus (dorsal scapular) nerve to the subclavius, and nerve to the serratus anterior muscle (long thoracic).

Within the limits of the posterior triangle of the neck, above the scalenus posterior muscle, dissect in succession the levator scapulae muscle and, superior to it, the splenius capitis muscle.

Cut horizontally through the middle of the scalenus anterior muscle and reflect the parts upward and downward. Study the course and relationships of the subclavian artery behind the muscle.

Dissect the trunks of the brachial plexus from the intermuscular space between the scalenus anterior and medius

Table 7

Name	Origin	Insertion	Innervation	Action
Levator scapulae	C1 to C4 transverse processes	Medial border of scapula	Ventral rami of C3 to C5 nerves	Stabilizes, retracts and elevates scapula
Longus colli	C5 to T3 vertebral bodies and anterior tubercles of C3 to C5 transverse processes	Anterior tubercles of atlas and of C5 and C6 transverse processes and C2 to C4 vertebral bodies	Ventral rami of C2 to C6 nerves	Anteriorly and laterally flexes cervical column
Scalenus anterior	Anterior tubercles of C3 to C6 transverse processes	Scalene tubercle of first rib	Ventral rami of C4 to C6 nerves	Anteriorly and laterally flexes cervical column; elevates first rib
Scalenus medius	Transverse process of axis and posterior tubercles of C3 to C7 transverse processes	First rib	Ventral rami of C3 to C8 nerves	Laterally flexes cervical column; elevates first rib
Scalenus posterior	Posterior tubercles of C4 to C6 transverse processes	Second rib	Ventral rami of C6 to C8 nerves	Laterally flexes cervical column; elevates second rib
Splenius capitis	Ligamentum nuchae and spines of C7 to T3 vertebrae	Mastoid process; occipital bone just below the superior nuchal line	Dorsal rami of C3 to C5 nerves	Extends head; rotates face to side of action
Sternocleidomastoid (see Table 5)				
Sternohyoid (see Table 5)				

muscles toward the spinal cord and separate the ventral rami of the C5, C6, C7, C8, and T1 spinal nerves.

Observe the position of the pleural dome within the posterior triangle of the neck (p. 114).

Summary of structures

1. Arteries
   Ascending cervical
   Costocervical trunk
   Deep cervical
   Dorsal scapular
   Highest intercostal
   Inferior thyroid
   Internal thoracic
   Subclavian
   Suprascapular
   Thyrocervical trunk
   Transverse cervical
   Vertebral
2. Fasciae
   Dome of pleura
   Muscular
3. Gland
   Thyroid
4. Joint
   Sternoclavicular
5. Lymphatic vessel
   Thoracic duct
6. Muscles
   Levator scapulae
   Longus colli
   Scalenus anterior
   Scalenus medius
   Scalenus posterior
   Splenius capitis
   Sternocleidomastoid
   Sternohyoid
7. Nerves
   Great auricular
   Lesser occipital
   Nerve to rhomboideus muscles
   Nerve to serratus anterior muscle
   Nerve to subclavius muscle
   Phrenic
   Supraclavicular
   Sympathetic trunk
   Ventral ramus of C1
   Ventral ramus of C2
   Ventral ramus of C3
   Ventral ramus of C4
   Ventral ramus of C5
   Ventral ramus of C6
   Ventral ramus of C7
   Ventral ramus of C8
   Ventral ramus of T1
   Vertebral
8. Plexuses
   Brachial, branches
   Brachial, trunks
   Cervical, branches
9. Surface anatomy
   Posterior triangle of neck
10. Veins
   Internal jugular
   Subclavian
   Subclavian, tributaries

MUSCLES OF THE BACK OF THE NECK

Observation. Observe (p. 57) the spine of the C7 vertebra, the scapular spine, and the acromioclavicular joint.

Palpation. Locate by this technique (pp. 5, 21, and 57) the exact position of the external occipital protuberance, the superior nuchal line, and the spines from the C7 to T12 vertebrae. Palpate (p. 169) the scapular spine from the acromioclavicular joint laterally to the medial margin of the scapula medially. By moving the upper limb in abduction and adduction, localize the inferior angle of the scapula and from it the lateral margin.

Dissection. Incise the skin (Fig. 3) along the line extending from the vertex, over the external occipital protuberance, to the spine of the T12 vertebra. Reflect the skin laterally and inferiorly as far as possible.

Dissect the occipitalis muscle, which is a subcutaneous muscle, in the subcutaneous layer of the occipital region from the superior nuchal line superiorly. Observe how the muscle gradually becomes continuous with the galea aponeurotica.

At a point approximately 1 cm lateral to and below the external occipital protuberance (p. 58), find the greater occipital nerve emeging from the aponeurosis of the trapezius muscle. Follow the nerve superiorly and identify the occipital artery, which approaches the nerve in this area.

Expose the entire surface of the trapezius muscle. Observe the passage through this muscle of several posterior rami of the regional spinal nerves. Cut the muscle vertically about 2 cm lateral to the median plane. Reflect the lateral

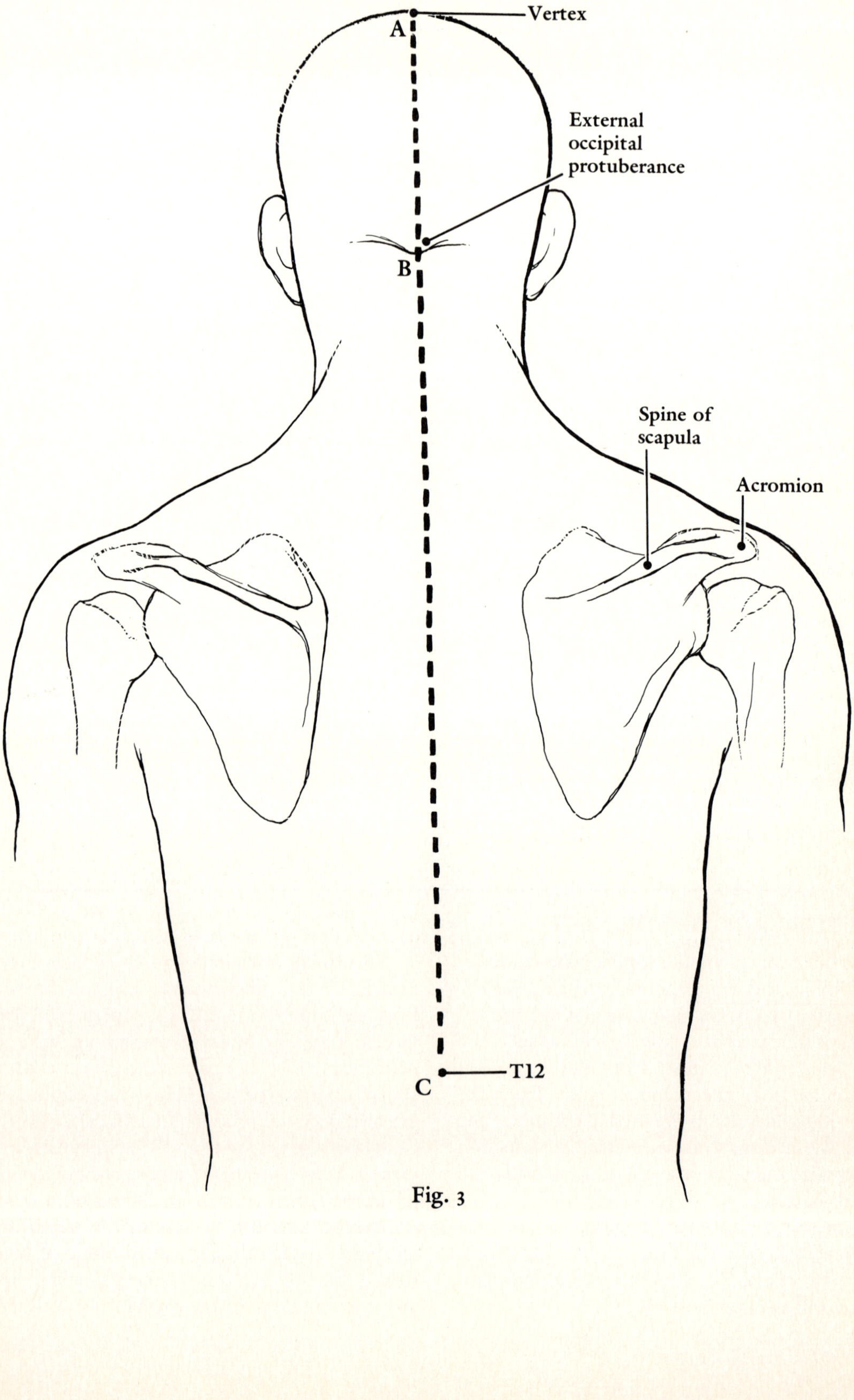
Vertex
A
External
occipital
protuberance
B
Spine of
scapula
Acromion
C
T12

Fig. 3

Table 8

Name	Origin	Insertion	Innervation	Action
Levator scapulae (see Table 7)				
Longissimus capitis	T1 to T4 transverse processes and C5 to C7 articular processes	Mastoid process	Dorsal rami of lower cervical, thoracic, and lumbar spinal nerves	Extends head; rotates face to side of action
Longissimus cervicis	T1 to T4 transverse processes	Posterior tubercles of C2 to C6 transverse processes	Dorsal rami of lower cervical, thoracic, and lumbar spinal nerves	Extends and laterally flexes cervical column
Multifidus	Articular processes	Spines	Dorsal rami of spinal nerves	Extends, laterally flexes and rotates vertebral column; posture
Obliquus capitis inferior	Spine of C2	Transverse process of C1	Dorsal ramus of C1 nerve	Rotates face to side of action
Obliquus capitis superior	Transverse process of C1	Between superior and inferior nuchal lines	Dorsal ramus of C1 nerve	Extends and laterally flexes head
Occipitalis	Highest nuchal line	Galea aponeurotica	Posterior auricular branch of facial nerve	Retracts scalp
Posterior superior serratus	Ligamentum nuchae, spines of C7 to T3, and supraspinous ligaments	Second to fifth ribs	Second to fifth intercostal nerves	Not known
Rectus capitis posterior major	Spine of C2	Inferior nuchal line	Dorsal ramus of C1	Extends head; rotates face to side of action
Rectus capitis posterior minor	Posterior tubercle of C1	Inferior nuchal line	Dorsal ramus of C1	Extends head
Rhomboideus major	T2 to T5 spines and supraspinous ligaments	Medial border of scapula	Dorsal scapular nerve	Retracts and rotates scapula
Rhomboideus minor	Ligamentum nuchae and C7 and T1 spines	Medial end of spine of scapula	Dorsal scapular nerve	Retracts and rotates scapula
Semispinalis capitis	C7 to T6 transverse processes	Between superior and inferior nuchal lines	Dorsal rami of cervical and thoracic spinal nerves	Extends head; rotates face to side opposite action
Semispinalis cervicis	T1 to T5 transverse processes	Spines of C2 to C5	Dorsal rami of cervical and thoracic spinal nerves	Extends cervical column; rotates cervical column to side opposite action
Splenius capitis (see Table 7)				
Splenius cervicis	Spines of T3 to T6	Posterior tubercles of C1 and C2 transverse processes	Dorsal rami of lower cervical spinal nerves	Extends cervical column; rotates face to side of action
Trapezius (see Table 5)				

part of the muscle as far as the scapular spine while removing the medial part completely from the spines of C7 and thoracic vertebrae.

Deep to the trapezius muscle (p. 60), from the medial margin of the scapula toward the ligamentum nuchae and the spines of C7 and upper thoracic vertebrae, dissect the rhomboideus minor and major muscles. Cut these muscles vertically about 2 cm lateral to the median plane, and reflect the lateral part of the muscles onto the scapula while removing the medial parts completely from the spines. Observe on the deep aspect of the reflected muscular parts the dorsal scapular nerve and the deep branch of the transverse cervical artery.

Deep to the rhomboideus muscles, extending from ribs two, three, four, and five to the lower cervical and the upper thoracic spines, dissect and subsequently remove the posterior superior serratus muscle.

Expose the splenius capitis muscle from the ligamentum nuchae and thoracic vertebral spines up to the lateral part of the superior nuchal line. Dissect the splenius cervicis muscle from those vertebral spines below the level of the spinal attachment of the splenius capitis muscle to the upper four cervical transverse processes. The muscular bodies of the two splenius muscles are not readily distinguishable from one another.

Along the plane of the two splenius muscles and lateral to them, expose the posterior aspect of the levator scapulae muscle from the medial angle of the scapula up to the first four cervical transverse processes.

Define the triangular space, which is confined between the right and left splenius capitis muscles and the superior nuchal line. The greater occipital nerves pass through this area on their way from the deep muscular layer to the subcutaneous layer.

Remove completely the splenius capitis and splenius cervicis muscles and the four heads of the levator scapulae muscle from their attachments to the transverse processes.

Dissect, on the next muscular plane (p. 62), the semispinalis capitis muscle from the upper thoracic transverse processes up to the superior nuchal line. Notice its medial and lateral halves. Immediately lateral to the semispinalis capitis muscle observe the muscular bundles of the longissimus capitis and longissimus cervicis muscles. Notice the passage through the semispinalis capitis muscle of the greater occipital nerve. Taking care to preserve this nerve, remove the semispinalis capitis muscle and the two longissimus muscles.

Locate the tip of the C2 spine and, from it down to the transverse processes of the T2 and T3 vertebrae, dissect the semispinalis cervicis muscle (p. 64). Remove this muscle and observe the multifidus muscle underneath.

Dissect the muscles of the suboccipital triangle—the rectus capitis posterior major, the obliquus capitis superior, and the obliquus capitis inferior muscles. Expose within the triangle the posterior arch of the atlas and, above it, the vertebral artery and the C1 spinal nerve. The posterior arch is attached to the occipital bone by the atlantooccipital membrane and to the axis by the atlantoaxial membrane. Observe the relationship of the former membrane to the verteral artery. Trace the dorsal ramus of the C2 nerve, the greater occipital nerve, as it passes around and below the obliquus capitis inferior muscle to end at the spinal ganglion. Remove the rectus capitis posterior major muscle and expose the rectus capitis posterior minor underneath (p. 66).

Clear all the muscles of the back of the neck from the posterior arches, spines, and the transverse processes of the cervical vertebrae. Remove the posterior vertebral segments (peduncles, laminae, spines) with a chisel or saw. Demonstrate subsequently the cervical spinal dura mater in the vertebral canal (p. 68). Complete the dissection of the C1 to C8 spinal nerves toward the spinal cord and expose their ganglia by removing carefully the dural sheath (p. 70). Study the composition of the brachial and cervical plexuses.

Follow the vertebral artery through the transverse foramina (vertebrarterial foramina) of the cervical vertebrae by removing the posterior parts of these processes with a chisel or saw.

Summary of structures

1. Arteries
   - Occipital
   - Transverse cervical, deep branch
   - Vertebral
2. Bones and bony landmarks
   - External occipital protuberance
   - Occipital
   - Posterior arch of atlas
   - Ribs
   - Scapula
   - Scapula, inferior angle
   - Scapula, lateral margin
   - Scapula, medial angle
   - Scapula, medial margin
   - Spines of C2 and C7 to T12 vertebrae
   - Spine of scapula
   - Superior nuchal line
   - Transverse foramina (vertebrarterial foramina)
   - Transverse process of T2 vertebra
   - Transverse process of T3 vertebra
   - Vertebral canal
   - Vertebral laminae
   - Vertebral peduncles
   - Vertebral spines
   - Vertebral transverse processes
   - Vertex
3. Central nervous system tissue and dural structures
   - Cervical spinal dura
   - Dorsal root ganglia
   - Spinal cord

4. Fasciae
   Aponeurosis of trapezius muscle
   Galea aponeurotica
5. Joint
   Acromioclavicular
6. Ligament
   Nuchal
7. Membranes
   Atlantoaxial
   Atlantooccipital
8. Muscles
   Levator scapulae
   Longissimus capitis
   Longissimus cervicis
   Multifidus
   Obliquus capitis inferior
   Obliquus capitis superior
   Occipitalis
   Posterior superior serratus
   Rectus capitis posterior major
   Rectus capitis posterior minor
   Rhomboideus major
   Rhomboideus minor
   Semispinalis capitis
   Semispinalis cervicis
   Splenius capitis
   Splenius cervicis
   Trapezius
9. Nerves
   C1 to C8 spinal
   Dorsal ramus of C2 spinal
   Dorsal scapular
   Greater occipital
   Posterior (dorsal) rami
10. Plexuses
    Brachial
    Cervical
11. Regions
    Suboccipital triangles
12. Surface anatomy
    Posterior triangle of neck
    Occipital region

REMOVAL OF CALVARIA: THE CEREBRUM, CEREBELLUM, CRANIAL NERVES AND VESSELS, AND THE DURAL VENOUS SINUSES

Palpation. Locate (p. 6) the lambda, bregma, and the parietal eminence.

Dissection. Cut all soft tissues horizontally around the skull along a line about 1 cm above the superciliary arches and the external occipital protuberance. Remove first the scalp layers from the bones. Subsequently cut the skull all around, following the same horizontal line, and detach the skullcap from the rest of the skelton, taking care not to damage the underlying dura mater (p. 73). Observe the frontal air sinus on the horizontal section through the vertical plate of the frontal bone, as well as the middle meningeal artery and its branches on the lateral surface of the dura mater.

Open with scissors the superior sagittal sinus from anterior as far posteriorly as the confluence of sinuses. Cut the dura mater on the right side horizontally, following the line of section of the scalp and skull from the midpoint of the forehead to the internal occipital protuberance, posteriorly. Reflect this part of the dura upward and simultaneously expose the surface of the right cerebral hemisphere (p. 74). Reflect now the entire right cerebral hemisphere laterally and observe the right profile of the falx cerebri, which is located between the two hemispheres. Reflect the posterior part of the brain—the occipital lobe—upward and observe the right half of the tentorium cerebelli. Gradually remove the right cerebral hemisphere and the right half of the brainstem above the tentorium cerebelli, separating them midsagittally from the left mass by severing those cranial nerves which arise from them.

Open the inferior sagittal and the straight sinuses through their right walls as far posteriorly as the confluence of sinuses (p. 75).

Starting from the cavernous sinus, open the superior petrosal sinus on the right side, lateral and posterior to the junction of this sinus with the transverse sinus. Subsequently open the transverse sinus from here medially, toward the confluence of sinuses.

Follow the same dissecting procedure on the left side.

Detach the falx cerebri from the crista galli and reflect it posteriorly. Observe the junction of the great cerebral vein with the inferior sagittal sinus, which represents the beginning of the straight sinus. Study all the attachments of the tentorium cerebelli to the skull bones (p. 76). Remove the tentorium cerebelli by cutting it close to the bony attachments, from in front backward, around, and then toward the confluence of sinuses.

Reflect the remaining portion of the brainstem posteriorly and then separate successively the cranial nerves that arise from it. Finally cut the brainstem horizontally at the level of the foramen magnum. Remove this part of the brainstem together with the cerebellum from the cranial cavity. Oberve the falx cerebelli between the two cerebellar hemispheres (p. 78).

Open the sigmoid sinus, proceeding from the junction between the transverse and superior petrosal sinuses toward the jugular foramen. From here follow the inferior petrosal sinus up to the cavernous sinus.

Observe and study the three cranial fossae.

Notice the olfactory bulb on the cribriform plate of the ethmoid bone (p. 78).

Expose both optic nerves and the chiasma on the base of the skull. Behind the chiasma, cut circularly through the diaphragma sellae and remove it with the hypophysis and the infundibulum.

On the tip of the petrous part of the temporal bone observe the roots of the trigeminal nerve entering the cavum trigeminale. Open the cavum, expose the trigeminal ganglion in it, and follow the branches toward their respective foramina.

Identify the oculomotor, trochlear, and abducent nerves before they penetrate the dural substance, and follow them through the cavernous sinus as far anteriorly as the superior orbital fissure. Dissect further in the cavernous sinus the internal carotid artery and the sympathetic plexus around it—the internal carotid plexus.

Expose in the internal acoustic meatus the facial nerve, the vestibulocochlear nerve, and the internal auditory artery.

In front of the terminal part of the sigmoid sinus, within the jugular foramen, identify from anterior to posterior the glossopharyngeal, vagus, and accessory nerves. Find the hypoglossal nerve in the hypoglossal canal.

The described dissecting procedure should be modified in certain specimens to remove from the cranial cavity the intact cerebrum, cerebellum, and brainstem in one block. The dural structures—falx cerebri and tentorium cerebelli—in this instance would be carefully detached from their bony attachments and removed, together with the central nervous system organs, from their bony cavities. Next separate the dural structures from the nervous tissue and observe and study the anatomical details on the lateral (p. 80), superior (p. 82), inferior (p. 83), and midsagittal sectional (p. 81) surfaces of the brain, cerebellum, and brainstem.

Ascertain the points of entry or exit of the 12 cranial nerves on the inferior aspect of the nervous system organs (p. 83).

Dissect the branches of and the communications between the internal carotid artery (anterior and middle cerebral arteries, anterior and posterior communicating branches) and the vertebral arterial system (spinal and cerebellar branches, internal auditory artery, basilar artery, posterior cerebral artery) (p. 83).

Summary of structures

1. Arteries and branches
   - Anterior cerebral
   - Anterior communicating
   - Basilar
   - Cerebellar branches
   - Internal auditory
   - Internal carotid
   - Middle cerebral
   - Middle meningeal
   - Posterior cerebral
   - Posterior communicating
   - Spinal branches
   - Vertebral
2. Bones and bony landmarks
   - Anterior cranial fossa
   - Bregma
   - Cranial cavity
   - Cribriform plate
   - Crista galli
   - Ethmoid
   - External occipital protuberance
   - Foramen magnum
   - Frontal air sinus
   - Hypoglossal canal
   - Internal acoustic meatus
   - Internal occipital protuberance
   - Jugular foramen
   - Lambda
   - Middle cranial fossa
   - Parietal eminence
   - Petrous part of temporal
   - Posterior cranial fossa
   - Superciliary arch
   - Superior orbital fissure
3. Central nervous system
   - Brainstem
   - Cerebellar hemisphere
   - Cerebellum
   - Cerebral hemisphere
   - Cerebrum
   - Hypophysis
   - Inferior surface (structures)
   - Infundibulum of hypophysis
   - Lateral surface (structures)
   - Medial surface (structures)
   - Midsagittal sectional surface (structures)
   - Occipital lobe
   - Superior surface (structures)

4. Dural structures and sinuses
   Cavernous sinus
   Cavum trigeminale
   Confluence of sinuses
   Diaphragma sellae
   Dura mater
   Falx cerebelli
   Falx cerebri
   Inferior petrosal sinus
   Inferior sagittal sinus
   Sigmoid sinus
   Straight sinus
   Superior petrosal sinus
   Superior sagittal sinus
   Tentorium cerebelli
   Transverse sinus
5. Nerves
   Abducent
   Accessory
   Facial
   Glossopharyngeal
   Hypoglossal
   Mandibular
   Maxillary
   Oculomotor
   Olfactory bulb
   Ophthalmic
   Optic
   Optic chiasma
   Trigeminal ganglion
   Trigeminal roots
   Trochlear
   Vagus
   Vestibulocochlear
6. Plexus
   Internal carotid
7. Vein
   Great cerebral

EAR

Observation. Observe the features of the auricle and its cartilages. In the cranial fossa (pp. 7 and 18), on the anterosuperior aspect of the petrous part of the temporal bone, observe the arcuate eminence. On the posterosuperior aspect of the petrous part note the internal acoustic meatus and the subarcuate fossa.

Palpation. Palpate the cartilaginous framework of the auricle and the initial part of the cartilaginous external acoustic meatus. Ascertain the movements of the temporomandibular joint in a living subject with a finger in the external acoustic meatus.

Dissection. Dissection should be undertaken from above, through the cranial fossa, and subsequently from the lateral side.

Dissection from above. Cut through the dura mater all along the petrosquamous junction as far anterior as the foramen ovale and also along the superior petrosal sinus and reflect the dura from the anterosuperior aspect of the petrous part anteriorly. Cut the dura around and close to the opening of the internal acoustic meatus and remove it from the posterosuperior aspect of the petrous part.

On the anterosuperior aspect (p. 84) find the hiatus for the greater petrosal nerve and the nerve itself. Immediately lateral and anterior to this nerve, locate the lesser petrosal nerve. Dissect with a chisel the greater petrosal nerve while tracing it backward through the bone as far as the geniculate ganglion. Expose from the ganglion (p. 86) the beginning of the second intraosseous part of the facial nerve, which extends further posteriorly. Expose now the first intraosseous part of the facial nerve by removing first the roof of the facial canal and then the roof of the internal acoustic meatus.

Dissect in the meatus, beneath the facial nerve, the components of the vestibulocochlear nerve. The cochlear division is in front; upper and lower divisions of the vestibular nerve are behind.

Remove the bone, layer by layer, from the arcuate eminence and open the anterior semicircular canal from one end to the other. Observe the position of the lateral and posterior semicircular canals and then open them.

Anterior and lateral to the lateral semicircular canal remove the roofs of the tympanic cavity, the tegmen tympani, and the mastoid antrum. Observe the ossicles, chorda tympani, and the tendon of the tensor tympani muscle in the tympanic cavity (p. 86).

In front of the exposed internal acoustic meatus (p. 88) remove the bone, layer by layer, until the cochlea is identified. Carefully remove the superior half of the cochlear bony envelop while preserving the structures within. Study the modiolus, the osseous spiral lamina, the helicotrema, the scala tympani, the scala vestibuli, and the cochlear duct. Remove the upper part of the spiral lamina and open the modiolus. Note the communication, through which the cochlear nerve passes, between the modiolus and the internal acoustic meatus.

Dissection from the lateral side. Remove the auricle after the cartilaginous details have been ascertained (p. 90). Expose the lateral aspect of the mastoid process in relation to the external acoustic meatus. Open the mastoid antrum from the lateral side by removing the bone piecemeal through the mastoid process, just behind the external acoustic meatus (p. 92). Observe the mastoid air cells, the antrum, and the aditus ad antrum.

Remove the bone all around the external acoustic meatus to about the depth of the tympanic membrane. Study the lateral aspect of the membrane, its different parts, and its attachments. Separate the membrane from the bone all around and detach it from the malleus (p. 94).

Observe the ossicles, the chorda tympani, and the anterior ligament of the malleus from the lateral side. Dissect the chorda tympani backward, through the bone, toward its origin at the facial nerve.

Disarticulate the malleus from the incus. Remove the attachment of the tensor tympani muscle and the lateral ligament from the malleus. Cut the chorda tympani as far anteriorly as possible and reflect it backward. Disarticulate the incus from the stapes and remove it. Observe the position of the stapes in the fenestra vestibuli (oval window) and the attachment of the tendon of the stapedius muscle to the stapes.

Observe once again the geniculate ganglion from the lateral side and expose the second intraosseous part of the facial nerve by removing the bone. Note the second genu of the facial nerve and dissect from it the third intraosseous part of the nerve downward through the mastoid part of the temporal bone, as far as the stylomastoid foramen.

Remove the bone from the lateral aspect of the stapedius muscle.

Note (p. 96), on the promontory underneath the mucous membrane, the terminal part of the tympanic nerve, the tympanic plexus, and the lesser petrosal nerve. Observe the beginning of the bony auditory tube just in front of the promontory. On the posteroinferior aspect of the promontory locate the fenestra cochleae (round window) and the secondary tympanic membrane within it. Above and parallel to the auditory tube, expose the tensor tympani muscle by removing the lateral bony wall of its semicanal. Remove the muscle.

Remove the bone of the promontory by using a chisel; open the promontory and study its relationship with the cochlea.

Remove the semicircular canals. Open the vestibule and study the structures in it. Observe vestibular connections with the internal acoustic meatus, through which the divisions of the vestibular nerve pass.

Remove the internal wall of the auditory tube and demonstrate the internal carotid artery in the carotid canal. Expose the artery as far as the foramen lacerum.

Summary of structures

1. Artery
   Internal carotid
2. Bones and bony landmarks
   Aditus ad antrum
   Anterior semicircular canal
   Arcuate eminence
   Carotid canal
   Cochlea
   Cochlear duct
   Cranial fossa
   External acoustic meatus
   Facial canal
   Fenestra cochleae
   Fenestra vestibuli
   Foramen lacerum
   Foramen ovale
   Helicotrema
   Incus
   Internal acoustic meatus
   Lateral semicircular canal
   Malleus
   Mastoid air cells
   Mastoid antrum
   Modiolus
   Osseous spiral lamina
   Ossicles
   Petrosquamous fissure
   Posterior semicircular canal

Table 9

Name	Origin	Insertion	Innervation	Action
Stapedius	Wall of pyramidal eminence	Neck of stapes	Facial nerve	Opposes tensor tympani to push stapes into fenestra vestibuli; protects against high-intensity sound
Tensor tympani	Cartilage of auditory tube, greater wing of sphenoid, and wall of canal of tensor tympani	Handle of malleus	Branch of nerve to medial pterygoid muscle	Tenses tympanic membrane; protects against high-intensity sound

Promontory
Scala tympani
Scala vestibuli
Stapes
Stylomastoid foramen
Subarcuate fossa
Tegmen tympani
Temporal, mastoid process
Temporal, petrous part
Tympanic cavity
Vestibule

3. Cartilages
   Auricle
   External acoustic meatus
4. Dural structures
   Dura mater
   Superior petrosal sinus
5. Ganglion
   Geniculate
6. Joint
   Temporomandibular
7. Ligaments
   Anterior, of malleus
   Lateral, of malleus
8. Membranes
   Secondary tympanic
   Tympanic
9. Muscles
   Stapedius
   Tensor tympani
10. Nerves
   Chorda tympani
   Cochlear division
   Facial
   Facial, first intraosseous part
   Facial, second intraosseous part
   Facial, third intraosseous part
   Greater petrosal
   Lesser petrosal
   Tympanic
   Vestibular division
   Vestibulocochlear
11. Plexus
   Tympanic
12. Surface anatomy
   Auricle
   External acoustic meatus

ORBIT

Observation. Study the walls of the orbit (p. 2) and the communications between the orbit and the cranial cavity.

Palpation. Palpate the eyeball (p. 107) through the eyelids and the fatty tissue around the eyeball. With two fingers palpate the tarsal plates, first in the upper eyelid and then in the lower.

Dissection. The dissection should be performed successively from above and in front.

Dissection from above. Remove the dura mater from the entire bony roof of the orbit (p. 99).

Remove the posterior walls of the frontal sinuses and superior walls of the ethmoidal air cells.

Remove the bony roof of the orbit as well as that of the superior orbital fissure and the optic canal.

Observe the periorbita, through which the frontal nerve is usually seen. Cut the periorbita all along this nerve and remove it. Dissect the nerve backward toward its origin, and then forward as far as the supraorbital foramen (p. 100). Observe the bifurcation of this nerve into the supratrochlear nerve medially and the supraorbital nerve laterally.

Along the lateral orbital wall, immediately deep to the periorbita, dissect the lacrimal nerve as far anteriorly as the lacrimal gland and posteriorly to the ophthalmic nerve. Demonstrate the superior aspect of the lacrimal gland.

Along the medial orbital wall expose the upper border of the superior oblique muscle and the trochlear nerve, as the nerve penetrates into the posterior muscular part. Finish the dissection of the trochlear nerve posteriorly. Dissect the superior oblique muscle from its origin as far anteriorly as the trochlea.

Dissect the lateral rectus muscle from its origin to its insertion into the eyeball.

Dissect the upper aspect of the levator palpebrae superioris muscle from its origin toward its insertion within the upper eyelid. Note the aponeurotic extensions of this muscle laterally to the lacrimal fossa and medially to the trochlea. Cut the muscle across its middle and reflect the anterior part anteriorly. While reflecting the posterior part posteriorly, observe a branch of the upper division of the oculomotor nerve, which enters the muscle through its deep aspect.

Immediately deep to the reflected levator palpebrae superioris muscle expose the superior aspect of the superior rectus muscle from its origin to its insertion into the eyeball. Cut the muscle across the middle and reflect the anterior part anteriorly (p. 102). Deep to it the reflected tendon of the superior oblique muscle extends from the trochlea posterolaterally to become inserted into the eyeball. Reflect the posterior part of the superior rectus muscle posteriorly and observe on its deep muscular aspect the upper division of the oculomotor nerve.

Within the muscular cone, which is surrounded by the

rectus and oblique muscles, find the nasociliary nerve. Isolate the branches of this nerve: to the eyeball, the long ciliary nerves; the anterior and posterior ethmoidal nerves, both of which pass medially between the superior oblique and medial rectus muscles; and the infratrochlear nerve, which passes inferior to the trochlea to unite farther anteriorly with the supratrochlear nerve.

Cut the supraorbital nerve close to the corresponding foramen and the supratrochlear nerve peripheral to its junction with the infratrochlear nerve, and reflect them backward.

Observe the origin of the ophthalmic artery from the internal carotid and its close relationship with the optic nerve in the optic canal. Follow the artery through the orbit as it passes beneath, across the lateral aspect of, and then above the optic nerve. Display the collateral branches of the artery and simultaneously expose the optic nerve from the chiasma to the eyeball.

Lateral to the optic nerve and in front of that part of the ophthalmic artery which is on its lateral side, dissect the ciliary ganglion using fine forceps. Demonstrate the roots to the ganglion from the nasociliary and oculomotor nerves and the efferent branches—the short ciliary nerves to the eyeball.

Cut the superior oblique muscle close to the trochlea and reflect it posteriorly. Complete the dissection of the anterior and posterior ethmoidal nerves and vessels within their respective canals.

Dissect the medial and lateral rectus muscles from their origins to the eyeball. On the medial aspect of the lateral muscle observe the abducent nerve.

Cut the optic nerve (p. 104) close to the chiasma and by rotating the eyeball, reflect it anteriorly as much as possible.

From the inferior aspect of the reflected superior rectus muscle, follow the upper division of the oculomotor nerve backward until it joins the inferior division. In doing so, remove as much of the lateral and superior rectus muscles as is necessary. Complete the dissection of the oculomotor nerve: follow the inferior, lower division forward and demonstrate its branches to the medial rectus, the inferior rectus, the inferior oblique, and the ciliary ganglion.

Expose the superior aspect of the inferior rectus muscle from its origin to the eyeball.

Dissection from the front. Reflect the orbicularis oculi muscle from the periphery toward the palpebral fissure and remove it. Deep to the muscle, within the eyelids, observe the orbital septum, the superior and inferior tarsal plates, and the lateral and medial palpebral ligaments (p. 108). Note the attachments of the two ligaments to the tarsal plates. Remove the lateral palpebral ligament from the bone and reflect it, together with the tarsal plates and the medial palpebral ligament, medially toward the medial orbital margin. Examine the termination of the levator palpebrae superioris muscle in the upper lid. Demonstrate the lacrimal canals, the lacrimal sac, and the beginning of the nasolacrimal duct. Open the lacrimal sac with scissors.

Keep the conjunctiva intact where it reflects from the

Table 10

Name	Origin	Insertion	Innervation	Action
Inferior oblique	Orbital surface of maxilla	Inferolateral, posterior quadrant of sclera	Lower division of oculomotor nerve	Elevates, abducts, and laterally rotates pupil
Inferior rectus	Common annular tendon	Inferior aspect of anterior hemisphere of sclera	Lower division of oculomotor nerve	Depresses, adducts, and laterally rotates pupil
Lateral rectus	Common annular tendon	Lateral aspect of anterior hemisphere of sclera	Abducent nerve	Abducts pupil
Levator palpebrae superioris	Inferior surface of lesser wing	Upper tarsal plate and skin of upper eyelid	Upper division of oculomotor nerve	Elevates upper eyelid
Medial rectus	Common annular tendon	Medial aspect of anterior hemisphere of sclera	Lower division of oculomotor nerve	Adducts pupil
Orbicularis oculi (see Table 1)				
Superior oblique	Sphenoid body, superomedial to optic canal	Superolateral, posterior quadrant of sclera	Trochlear nerve	Depresses, abducts, and medially roates pupil
Superior rectus	Common annular tendon	Superior aspect of anterior hemisphere of sclera	Upper division of oculomotor nerve	Elevates, adducts, and medially rotates pupil

posterior aspect of the lids onto the eyeball. Remove the remainder of the orbital septum.

Dissect the eyeball all around, and demonstrate the attachments of the rectus muscles on it (p. 110).

Dissect the inferior oblique muscle from the medial part of the infraorbital margin laterally, upward, and backward. The muscle is usually located between the two lobes of fatty tissue. Remove the muscle from the bone.

Cut from above the inferior, lateral, and medial rectus muscles across their middle. Free the tendon of the superior oblique muscle from the trochlea. Remove the eyeball, together with the attached muscular parts and the optic nerve, from the orbit. Study the surface of the eyeball and the attachments on it of the muscles and the conjunctiva.

The two dissecting procedures could also be simultaneously performed on both sides of the body.

Summary of structures

1. Arteries
   Anterior ethmoidal
   Internal carotid
   Ophthalmic
   Posterior ethmoidal
2. Bones and bony landmarks
   Anterior ethmoidal canal
   Cranial cavity
   Lacrimal fossa
   Optic canal
   Orbit
   Posterior ethmoidal canal
   Superior orbital fissure
   Supraorbital foramen
3. Cartilage
   Tarsal
4. Dural structures
   Dura mater
   Periorbita
5. Ganglion
   Ciliary
6. Gland
   Lacrimal
7. Lacrimal apparatus
   Canal
   Nasolacrimal duct
   Sac
8. Ligaments
   Lateral palpebral
   Medial palpebral
9. Membranes
   Orbital septum
   Tendon of superior oblique muscle
   Trochlea
10. Muscles
   Inferior oblique
   Inferior rectus
   Lateral rectus
   Levator palpebrae superioris
   Medial rectus
   Orbicularis oculi
   Superior oblique
   Superior rectus
11. Nerves
   Abducent
   Anterior ethmoidal
   Frontal
   Infratrochlear
   Lacrimal
   Long ciliary
   Nasociliary
   Oculomotor
   Oculomotor, lower division
   Oculomotor, upper division
   Ophthalmic
   Optic
   Optic chiasma
   Posterior ethmoidal
   Short ciliary
   Supraorbital
   Supratrochlear
   Trochlear
12. Paranasal sinuses
   Ethmoidal cells
   Frontal
13. Surface anatomy
   Conjunctiva
   Eyeball
   Eyelid
   Palpebral fissure
14. Veins
   Anterior ethmoidal
   Posterior ethmoidal

PHARYNX AND PREVERTEBRAL REGION

Dissection. Turn the cadaver face downward and place wooden blocks beneath the chest.

Reflect the cervical part of the spinal cord posteriorly and inferiorly. Observe the posterior longitudinal ligament along the posterior aspect of the vertebral bodies (p. 112). Further dissect the cruciform, apical, and alar ligaments and the tectorial membrane of the median atlantoaxial joint.

Disarticulate both atlantooccipital joints and cut horizontally, at the level of these joints, ligaments and membranes and the rectus capitis lateralis and anterior muscles. Reflect next the entire head, together with the organs of the neck, anteriorly and inferiorly from the vertebral column and the prevertebral muscles (p. 114). To achieve this it is advisable to use the cleavage layer of the retropharyngeal space.

Turn the cadaver on its back, and on the anterior aspect of the vertebral column expose the longus capitis and longus colli muscles, and the anterior longitudinal ligament between them.

Expose the superior constrictor muscle of the pharynx from behind (p. 116) as far down as the upper border of the middle constrictor. Demonstrate the basipharyngeal membrane between the uppermost part of this muscle and the base of the skull, the pharyngeal tubercle (p. 8).

Dissect the middle pharyngeal constrictor muscle from its upper border as far down as the upper border of the inferior pharyngeal constrictor muscle, which overlaps it.

Dissect the inferior pharyngeal constrictor muscle from its upper border to the junction of the pharynx with the esophagus.

Proceeding from the neck superiorly toward the base of the skull, complete the dissection of the vagus nerve and the superior laryngeal branch, accessory nerve, hypoglossal nerve, the common and internal carotid arteries, the cervical sympathetic trunk and ganglia, and the internal jugular vein. Lateral to the superior pharyngeal constrictor muscle and at a more anterior plane to these structures, identify the styloid process, and from the processs dissect the stylopharyngeus muscle. Note the attachment of this muscle to the pharyngeal wall between the superior and middle pharyn-

Table 11

Name	Origin	Insertion	Innervation	Action
Inferior pharyngeal constrictor	Side of cricoid cartilage and oblique line of thyroid lamina	Median raphe	Pharyngeal plexus	Constricts pharynx
Longus capitis	Anterior tubercles of C3 to C6 transverse processes	Basilar part of occipital bone	Ventral rami of C1 to C3 nerves	Flexes head
Longus colli (see Table 7)				
Middle pharyngeal constrictor	Lesser and greater horns of hyoid bone and stylohyoid ligament	Median raphe	Pharyngeal plexus	Constricts pharynx
Palatopharyngeus	Posterior border of hard palate and palatine aponeurosis	Posterior border of thyroid cartilage	Pharyngeal plexus	Elevates pharyngeal wall
Rectus capitis anterior	Anterior surface of lateral mass and transverse process of C1	Anterior to occipital condyle	Ventral rami of C1 and C2 nerves	Flexes head
Rectus capitis lateralis	Upper surface of transverse process of C1	Jugular process of occipital bone	Ventral rami of C1 and C2 nerves	Laterally flexes head
Stylopharyngeus	Styloid process	Pharyngeal wall between superior and middle constrictor muscles	Glossopharyngeal nerve	Elevates pharynx
Superior pharyngeal constrictor	Pterygoid medial plate and hamulus, pterygomandibular raphe, mylohyoid line of mandible, side of tongue	Median raphe	Pharyngeal plexus	Constricts pharynx

geal constrictor muscles. During the dissection of the stylopharyngeus muscle take care to keep all structures posterior to it intact.

On the posterior aspect and around the inferior border of the stylopharyngeus muscle find the glossopharyngeal nerve and its continuation, the lingual branch. Dissect the nerve toward the jugular foramen. Isolate the branches of the glossopharyngeal, vagus, and accessory nerves to the pharyngeal musculature.

With scissors cut the pharynx along the raphe from the pharyngoesophageal junction superiorly. Remove the basipharyngeal membrane from the base of the skull and reflect both halves of the pharyngeal wall laterally (p. 118).

With the pharynx open, observe the mucous membrane in three chambers—the nasopharynx, oropharynx, and laryngopharynx.

In the nasopharynx observe the choanae, the pharyngeal opening of the auditory tube, the torus tubarius, the pharyngeal recess, and the soft palate.

In the oropharynx study the faucial isthmus, the uvula, and the palatopharyngeal fold. Cut the mucous membrane along the palatopharyngeal fold and expose the palatopharyngeus muscle in it.

In the laryngopharynx study the glossoepiglottic valleculae, the median and lateral glossoepiglottic folds, the epiglottic cartilage, the aryepiglottic folds, the tips of the arytenoid cartilages, and the interarytenoid fold. Below the level of the laryngeal inlet observe from lateral to medial the posterior borders of the thyroid cartilage, the piriform recesses, and the relief of the plate of the cricoid cartilage.

Summary of structures

1. Arteries
   - Common carotid
   - Internal carotid
2. Bones and bony landmarks
   - Base of skull
   - Jugular foramen
   - Pharyngeal tubercle
   - Styloid process
   - Vertebral bodies
   - Vertebral column
3. Cartilages
   - Arytenoid
   - Cricoid
   - Epiglottic
   - Thyroid
4. Central nervous system
   - Cervical spinal cord
5. Folds
   - Aryepiglottic
   - Interarytenoid
   - Lateral glossoepiglottic
   - Median glossoepiglottic
6. Ganglion
   - Cervical sympathetic
7. Joints
   - Atlantooccipital
   - Median atlantoaxial
8. Ligaments
   - Alar
   - Anterior longitudinal
   - Apical
   - Cruciform
   - Posterior longitudinal
9. Membranes
   - Basipharyngeal
   - Tectorial
10. Muscles
    - Inferior pharyngeal constrictor
    - Longus capitis
    - Longus colli
    - Middle pharyngeal constrictor
    - Palatopharyngeus
    - Prevertebral
    - Raphe
    - Rectus capitis anterior
    - Rectus capitis lateralis
    - Stylopharyngeus
    - Superior pharyngeal constrictor
11. Nerves
    - Accessory
    - Glossopharyngeal
    - Hypoglossal
    - Lingual of glossopharyngeal
    - Pharyngeal of accessory
    - Pharyngeal of glossopharyngeal
    - Pharyngeal of vagus
    - Superior laryngeal
    - Sympathetic trunk
    - Vagus
12. Organs
    - Esophagus
    - Laryngeal inlet
    - Pharyngoesophageal junction
    - Pharynx
    - Soft palate
    - Torus tubarius
    - Uvula
13. Spaces
    - Choana
    - Faucial isthmus
    - Glossoepiglottic vallecula
    - Laryngopharynx
    - Nasopharynx
    - Oropharynx
    - Pharyngeal opening of auditory tube
    - Pharyngeal recess
    - Piriform recess
    - Retropharyngeal
14. Vein
    - Internal jugular

NASAL CAVITY, PARANASAL SINUSES, AND PTERYGOPALATINE FOSSA

Observation. Study the walls of the nasal cavity (p. 9) and the pterygopalatine fossa in the skull.

Palpation. Palpate the external features of the nose (p. 120) and the cartilages (p. 121): dorsum nasi, ala nasi, external orifice, septum, piriform aperture, and the septal, lateral and alar cartilages.

Dissection. This procedure is divided into the dissection of the nasal cavity and paranasal sinuses and the dissection of the pterygopalatine fossa.

Nasal cavity and paranasal sinuses. With a saw cut the skull, including the mandible and the tongue, at about 0.5 cm to the left of the median plane. Continue sectioning inferiorly to the level of the epiglottis and the hyoid bone. Separate the two segments of the head and study the nasal septum as a whole on the right section. After the mucosa has been removed from the left face of the nasal septum (p. 122), the septal components, the vomer, the perpendicular plate of the ethmoid bone, and the septal cartilage should be observed. Remove the nasal septum from the roof and floor of the nasal cavity and study the lateral nasal wall. Observe three, occasionally four, nasal conchae on the lateral nasal wall (p. 124).

Remove the inferior nasal concha (p. 126) from the lateral nasal wall and find the nasal opening of the nasolacrimal duct underneath it.

Remove the middle nasal concha from the lateral nasal wall and demonstrate the ethmoidal bulla, uncinate process, ethmoid infundibulum, and the openings of the frontal anterior ethmoidal (not observable because of small size) and maxillary sinuses.

Remove the superior nasal concha and ascertain (not observable because of their small size) the area where the posterior ethmoidal air cells open.

Within the sphenoethmoidal recess, the space superior to the uppermost concha, find the opening of the sphenoidal sinus. Examine this sinus also on the sagittal section.

Remove the mucous membrane from the posterior part of the lateral nasal wall and from the perpendicular plate of the palatine bone and expose the sphenopalatine foramen (p. 128).

Open the maxillary sinus from the nasal cavity.

Pterygopalatine fossa. Dissect the maxillary nerve from the trigeminal ganglion as far as the foramen rotundum. Remove the upper bony boundary of the foramen and the roof of the pterygopalatine fossa. Dissect, in and from the pterygopalatine fossa, the maxillary nerve and its continuation, the infraorbital nerve, through the inferior orbital fissure while removing the roof of the fissure. Follow the nerve farther anteriorly through the infraorbital groove and the canal until it passes into the face at the infraorbital foramen. Note the infraorbital artery, a branch of the maxillary artery, which accompanies the nerve. During this procedure isolate the branches of either the maxillary or the infraorbital nerve—posterior, middle and anterior superior alveolar, and zygomatic.

Locate the sphenopalatine foramen in the nasal cavity and, from the foramen inferiorly, remove all of the perpendicular plate of the palatine bone (p. 130). Expose, in the pterygopalatine fossa, the pterygopalatine (sphenopalatine) ganglion and the greater and lesser palatine nerves. Note the descending palatine artery, a branch of the maxillary artery, which courses together with the palatine nerves. Dissect the artery and the nerves down to the level of the hard palate.

Dissect the nerve of the pterygoid canal (p. 133) from the pterygopalatine (sphenopalatine) ganglion posteriorly, through the root of the pterygoid process, toward the foramen lacerum, where the greater and deep petrosal nerves unite to form the nerve of the pterygoid canal.

Incise the mucous membrane from the pharyngeal opening of the auditory tube to the soft palate and underneath this mucosa demonstrate the levator veli palatini muscle. After this muscle has been removed, expose at a more anterior and lateral plane the tensor veli palatini muscle, which extends vertically along the medial pterygoid plate. Verify how the tendon of the latter muscle curves around the

Table 12

Name	Origin	Insertion	Innervation	Action
Levator veli palatini	Petrous part and vaginal process of tympanic part of temporal bone, cartilage of auditory tube	Palatine aponeurosis	Pharyngeal plexus	Elevates soft palate
Tensor veli palatini	Scaphoid fossa of pterygoid process, cartilage of auditory tube, spine of sphenoid bone	Palatine aponeurosis	Mandibular nerve	Tenses soft palate

hamulus of the pterygoid process before it penetrates horizontally into the substance of the soft palate.

Open all of the paranasal sinuses from the nasal cavity (p. 134). Observe on the roof of the maxillary sinus the projection of the infraorbital nerve and the blood vessels contained in the groove and canal. If the sphenoidal sinus is well developed, in particular its pterygoid recess, the projection of the pterygoid canal in it can also be observed.

Summary of structures

1. Arteries
   Descending palatine
   Infraorbital
   Maxillary
2. Bones and bony landmarks
   Foramen lacerum
   Foramen rotundum
   Hamulus
   Hyoid
   Inferior orbital fissure
   Infraorbital canal
   Infraorbital foramen
   Infraorbital groove
   Mandible
   Medial pterygoid plate
   Perpendicular plate of ethmoid
   Perpendicular plate of palatine
   Piriform aperture
   Pterygoid canal
   Pterygoid process
   Skull
   Sphenopalatine foramen
   Vomer
3. Cartilages
   Alar
   Epiglottis
   Lateral
   Septal
   Triangular
4. Ganglia
   Pterygopalatine (sphenopalatine)
   Trigeminal
5. Muscles
   Levator veli palatini
   Tensor veli palatini
6. Nasal structures
   Concha, inferior
   Concha, middle
   Concha, superior
   Ethmoid infundibulum
   Ethmoidal bulla
   Hard palate
   Maxillary sinus
   Openings of anterior ethmoidal air cells
   Opening of frontal sinus
   Opening of maxillary sinus
   Opening of nasolacrimal duct
   Openings of posterior ethmoidal air cells
   Opening of sphenoidal sinus
   Septum
   Sphenoidal sinus
   Sphenoidal sinus, pterygoid recess
   Uncinate process
7. Nerves
   Anterior superior alveolar
   Deep petrosal
   Greater palatine
   Greater petrosal
   Infraorbital
   Lesser palatine
   Maxillary
   Middle superior alveolar
   Posterior superior alveolar
   Of pterygoid canal
   Zygomatic
8. Other structures
   Pharyngeal opening of auditory tube
   Soft palate
   Tongue
9. Spaces
   Nasal cavity
   Paranasal sinuses
   Pterygopalatine fossa
   Sphenoethmoidal recess
10. Surface anatomy
   Ala nasi
   Dorsum nasi
   External nasal orifice
   Nose

TEETH, TONGUE, MOUTH, PALATINE TONSIL, AND PALATE

Dissection

Teeth. Observe the different types of teeth, if they are still preserved: incisors, canines, premolars, and molars. Study the occlusal surfaces of both the upper and the lower teeth.

Tongue. On the median sagittal section (p. 136) observe the superior and inferior longitudinal, vertical, and the transverse muscles, the intrinsic muscles, and the attachments of the tongue to the hyoid bone, the mental spine, and the epiglottic cartilage. Identify the gustatory papillae—the filiform, fungiform, foliate, and vallate—on the superior and lateral surfaces of the tongue. Study the lingual tonsil (p. 146).

Mouth. Examine the walls of the oral vestibule, the oral cavity proper, and the oral opening into the pharynx, the faucial isthmus. Find the vestibular opening of the parotid duct opposite the maxillary second molar. Reflect the tip of the tongue backward, and beneath it observe the opening of the submandibular duct at the sublingual papilla (p. 143). Remove the mucous membrane around this opening while following the submandibular duct posteriorly and simultaneously exposing the superior aspect of the sublingual gland (p. 144).

Display the lingual nerve and the submandibular duct under cover of the mucous membrane of the oral floor by removing the latter membrane from between the tongue and the alveolar process, the linguoalveolar groove. Note the relationship of the nerve to the duct at different anteroposterior levels. Follow the nerve posteriorly and identify its submandibular ganglion at the level where the nerve is closely apposed to the deep lobe of the submandibular gland.

Palatine tonsil. By reflecting the corresponding half of the tongue medioinferiorly (p. 138), observe the tonsillar fossa, which is bounded by the palatopharyngeal fold posteriorly, the palatoglossal fold anteriorly, and the tongue inferiorly. Identify and study the palatine tonsil within the fossa.

Incise the mucous membrane along the palatoglossal fold and expose, within the fold, the palatoglossus muscle. From the point at which the palatoglossus and palatopharyngeus muscles and folds meet (p. 140) on the soft palate, reflect the mucous membrane together with the palatine tonsil downward. Lateral to the reflected tonsil demonstrate the superior pharyngeal constrictor muscle and the tonsillar blood vessels, which pass through this muscle on their way to the tonsil. Remove the tonsil. Observe the insertion into the tongue of the constrictor muscle and the course of the lingual branch of the glossopharyngeal nerve.

Palate. With a needle pointed upward through the mucous membrane of the hard palate, find the greater palatine foramen medial to the maxillary second molar. Introduce the needle through the foramen into the pterygopalatine fossa.

Subsequently identify, behind and between the maxillary medial incisors, the incisive fossa, which communicates between the nasal and the oral cavities.

Remove the mucous membrane from the greater pala-

Table 13

Name	Origin	Insertion	Innervation	Action
Inferior longitudinal fibers	Submucosa of root of tongue	Submucosa of apex of tongue	Hypoglossal nerve	Shortens tongue
Palatoglossus	Palatine aponeurosis	Submucosa of side of tongue	Pharyngeal plexus	Elevates root of tongue
Palatopharyngeus (see Table 11)				
Superior longitudinal fibers	Submucosa close to epiglottis, median fibrous septum	Submucosa of edge of tongue	Hypoglossal nerve	Shortens tongue
Superior pharyngeal constrictor (see Table 11)				
Transverse fibers	Median fibrous septum	Submucosa of side of tongue	Hypoglossal nerve	Narrows and elongates tongue
Vertical fibers	Submucosa of dorsal surface of tongue	Ventral side of tongue	Hypoglossal nerve	Flattens and broadens tongue

tine foramen and dissect the branches of both the greater and the lesser palatine nerves anteriorly and posteriorly.

Remove the mucous membrane from the incisive fossa and dissect the anterior palatine nerve, a branch of the nasopalatine nerve, posteriorly through the mucosa of the hard palate. Note the connections between the anterior and greater palatine nerves.

Summary of structures

1. Artery
   Tonsilar
2. Bones and bony landmarks
   Alveolar process
   Greater palatine foramen
   Hyoid
   Incisive foramen
   Mental spine
   Pterygopalatine fossa
3. Cartilage
   Epiglottic
4. Folds
   Palatoglossal
   Palatopharyngeal
5. Ganglion
   Submandibular
6. Glands
   Sublingual
   Submandibular, deep lobe
7. Gustatory papillae
   Filiform
   Foliate
   Fungiform
   Vallate
8. Muscles
   Inferior longitudinal fibers
   Intrinsic lingual
   Palatoglossus
   Palatopharyngeus
   Superior longitudinal fibers
   Superior pharyngeal constrictor
   Transverse fibers
   Vertical fibers
9. Nerves
   Anterior palatine
   Glossopharyngeal, lingual branch
   Greater palatine
   Lesser palatine
   Lingual
   Nasopalatine
10. Openings
    Oral, of submandibular duct (sublingual papilla)
    Vestibular, of parotid duct
11. Organs
    Canines
    Hard palate
    Incisors
    Molars
    Occlusal surfaces
    Premolars
    Soft palate
    Tongue
12. Regions
    Faucial isthmus
    Mouth
    Oral cavity
    Oral vestibule
    Pharynx
    Tonsillar fossa
13. Tonsils
    Lingual
    Palatine
14. Vein
    Tonsilar

LARYNX

Dissection. Remove all the infrahyoid and pharyngeal muscles from the lateral aspect of the thyroid cartilage and all the muscles from the hyoid bone.

Clean the surface of the thyrohyoid membrane and note how the internal branch of the superior laryngeal nerve and the superior laryngeal blood vessels pass through it.

Dissect the cricothyroid muscle and its nerve, the external branch of the superior laryngeal nerve (p. 150).

Locate the cricothyroid joints and, immediately posterior to them, under cover of the pharyngeal mucous membrane, expose the recurrent (inferior) laryngeal nerve (p. 148). Follow the nerve superiorly by removing the mucous membrane of the piriform recess and note its connections with the internal branch of the superior laryngeal nerve.

Proceed by removing the entire mucous membrane from the posterior aspect of the cricoid lamina; then expose the posterior cricoarytenoid muscles on both sides as far as the muscular process of the arytenoid cartilages.

Continue by removing the mucous membrane from the posterior aspects of the arytenoid cartilages and expose simultaneously the transverse and oblique parts of the interarytenoid muscle, which lies underneath the mucosa.

Remove the cricothyroid muscle on one side and disarticulate the cricothyroid joint on the same side. Remove the thyrohyoid membrane from the upper border and the superior horn of the thyroid cartilage. Cut the thyroid lamina vertically, off the median plane by about 2 or 3 mm, and remove it (p. 152). Between the cricoid arch and the muscular process of the arytenoid dissect the lateral cricoarytenoid muscle; immediately superior to it and on the same plane dissect the diffuse fibers of the thyroarytenoid muscle.

Locate the position of the vocal process of the arytenoid cartilage and expose the vocal ligament, proceeding horizontally and anteriorly from this process. Immediately superior to this ligament (p. 154) observe the diverticulum of the laryngeal mucous membrane, which is limited superiorly by the vestibular ligament. Excise the diverticulum between the two ligaments and thus open the middle part of the laryngeal cavity—the ventricle.

Remove, on the same side, the muscles from their cartilaginous attachments. Underneath the lateral cricoarytenoid and the thyroarytenoid expose the quandrangular membrane, which extends between the vestibular ligament, the epiglottic cartilage, the aryepiglottic fold, and the arytenoid cartilage. Expose also the triangular membrane—the conus elasticus—which extends between the vocal ligament, the superior border of the cricoid cartilage, and the internal aspect of the thyroid cartilage and the cricothyroid ligament. Remove the ligaments and the membranes from their cartilaginous attachments. Study the mucous membrane of the larynx on the opposite side (pp. 156 and 157). Remove the epiglottic cartilage from the thyroid cartilage and define both its surfaces. Disarticulate the cricoarytenoid joint and define the surfaces and the processes of the arytenoid cartilage. Demonstrate the movements of the cricoarytenoid joint. Remove the arytenoid cartilage.

Disarticulate the cricothyroid joint and remove the muscles so that the internal aspect of the thyroid lamina is exposed. Remove, define, and subsequently study the cricoid cartilage.

If the laryngeal cartilages have been dissected in situ, study their morphology and relationships (pp. 158 to 161). Also ascertain the relationship of the larynx as a whole with the other cervical structures (p. 162).

Summary of structures

1. Artery
   Superior laryngeal
2. Bone
   Hyoid
3. Cartilages
   Arytenoid
   Cricoid
   Cricoid arch
   Cricoid lamina
   Epiglottic
   Lamina of thyroid
   Laryngeal
   Muscular process of arytenoid
   Superior horn of thyroid
   Thyroid
   Vocal process of arytenoid
4. Fold
   Aryepiglottic

Table 14

Name	Origin	Insertion	Innervation	Action
Cricothyroid (see Table 6)				
Interarytenoid	Muscular process of arytenoid	Muscular process and apex of arytenoid and aryepiglottic fold on opposite side	Recurrent laryngeal nerve	Adducts vocal cords; sphincter of laryngeal inlet
Lateral cricoarytenoid	Arch of cricoid	Muscular process of arytenoid	Recurrent laryngeal nerve	Adducts vocal cords
Posterior cricoarytenoid	Posterior surface of cricoid lamina	Muscular process of arytenoid	Recurrent laryngeal nerve	Abducts and tenses vocal cords
Thyroartenoid	Internal surface of thyroid angle and cricothyroid ligaments	Anterolateral surface of arytenoid	Recurrent laryngeal nerve	Adducts vocal cords; shortens and relaxes vocal cords

5. Joints
   Cricoarytenoid
   Cricothyroid
6. Ligaments
   Cricothyroid
   Vestibular
   Vocal
7. Membranes
   Quadrangular
   Thyrohyoid
   Triangular
8. Muscles
   Cricothyroid
   Infrahyoid
   Interarytenoid
   Lateral cricoarytenoid
   Pharyngeal
   Posterior cricoarytenoid
   Thyroarytenoid
9. Nerves
   Recurrent (inferior) laryngeal
   Superior laryngeal, external branch
   Superior laryngeal, internal branch
10. Spaces
   Piriform recess
   Ventricle
11. Vein
   Superior laryngeal

Part Two
UPPER LIMB

PECTORAL REGION AND AXILLA

Observation. Observe (pp. 180 and 181) the mammary gland, areola, nipple, sternoclavicular and acromioclavicular joints, clavicle, jugular notch, sternum, sternal angle, ribs, and the medial and lateral condyles of the humerus.

Palpation. Palpate the spine of the scapula, acromion, acromioclavicular joint, clavicle, tip of the coracoid process, sternoclavicular joint, jugular notch, manubrium sterni, sternal angle, body of the sternum, xiphoid process, ribs, greater and lesser tubercles and borders of the humerus, medial and lateral epicondyles (pp. 168 to 173), anterior musculature of the arm, the deltoid muscle, and the anterior aspect of the pectoralis major muscle, particularly the inferior free border, at the level where the muscle extends from thorax to arm. To facilitate the palpation of the latter border, the arm should be abducted. Palpate, posterior to the pectoralis major muscle, the free border of the latissimus dorsi muscle at the level where it extends from the posterior aspect of the thorax to the arm. Observe the base of the axilla, which is limited by the free border of the pectoralis muscle anteriorly, the thorax medially, the free border of the latissimus dorsi muscle posteriorly, and the arm laterally. Then palpate the axillary content through the base of the axilla.

Dissection. Cut the skin (Fig. 4) starting from the acromioclavicular joint, along the lateral aspect of the arm as far as the lateral epicondyle of the humerus. From here continue the skin incision horizontally toward the medial epicondyle. Cut the skin vertically, following the anterior median line of the thorax, between the jugular notch and the xiphoid process and proceed from this point horizontally and backward as far as possible.

Reflect the epidermis and the dermis from the thorax laterally and from the arm medially.

In the subcutaneous tissue on the lateral aspect of the arm expose the cephalic vein (p. 182) and follow it toward the clavicle.

In a female cadaver (p. 182) dissect the mammary gland and identify the lobes, septa, lactiferous ducts, areola, and the nipple.

Lateral to the cephalic vein dissect the anterior portion of the deltoid muscle, and medial to the vein dissect the entire surface of the pectoralis major muscle. Cut the latter muscle vertically, 3 to 4 cm medial to the humeral attachment. Reflect the parts of the muscle medially and laterally and observe the lateral and medial pectoral nerves and blood vessels as they enter the muscle through its deep aspect.

Dissect the pectoralis minor muscle (p. 184). Between the anterosuperior border of this muscle and the clavicle dissect the branches of the thoracoacromial artery as they pass through the clavipectoral fascia. Observe along the same plane the fascial layer, which extends from the inferoposterior border of the pectoralis minor muscle to the arm. Separate by hand the muscle from the underlying axillary contents. Cut the muscle obliquely through the middle and reflect the parts to their respective attachments. In the course of the latter procedure verify the relationship of the muscle to the lateral and medial pectoral nerves and blood vessels.

Finish the dissection of the cephalic vein to its junction with the axillary vein (p. 186). Subsequently expose the axillary vein in the axilla while removing all of its tributaries.

Complete the dissection of the thoracoacromial artery as far as the origin from the axillary artery. Subsequently dissect the axillary artery (taking care to preserve the surrounding cords of the brachial plexus) up to the clavicle and down to the level of the inferior, free border of the pectoralis major muscle. Expose the collateral branches of the artery from superior to inferior: the superior thoracic, thoracoacromial, lateral thoracic, subscapular, and anterior and posterior circumflex humeral.

Dissect around the axillary artery, while preserving all of the dissected arteries and the vein, the lateral, medial, and posterior cords of the brachial plexus. Expose from the cords into the arm the initial parts of the terminal branches from the brachial plexus: the median, ulnar, medial antebrachial cutaneous, medial brachial cutaneous (which sometimes communicates with the second intercostal nerve [intercostobrachial]), musculocutaneous, radial, and axillary nerves. Note the relationship of the axillary artery with the cords of the plexus.

Dissect on the lateral aspect of the thorax (p. 188) the serratus anterior muscle. Note the long thoracic nerve on the surface of the muscle.

Cut the fascia in between the clavicle and the first rib horizontally and expose the subclavius muscle.

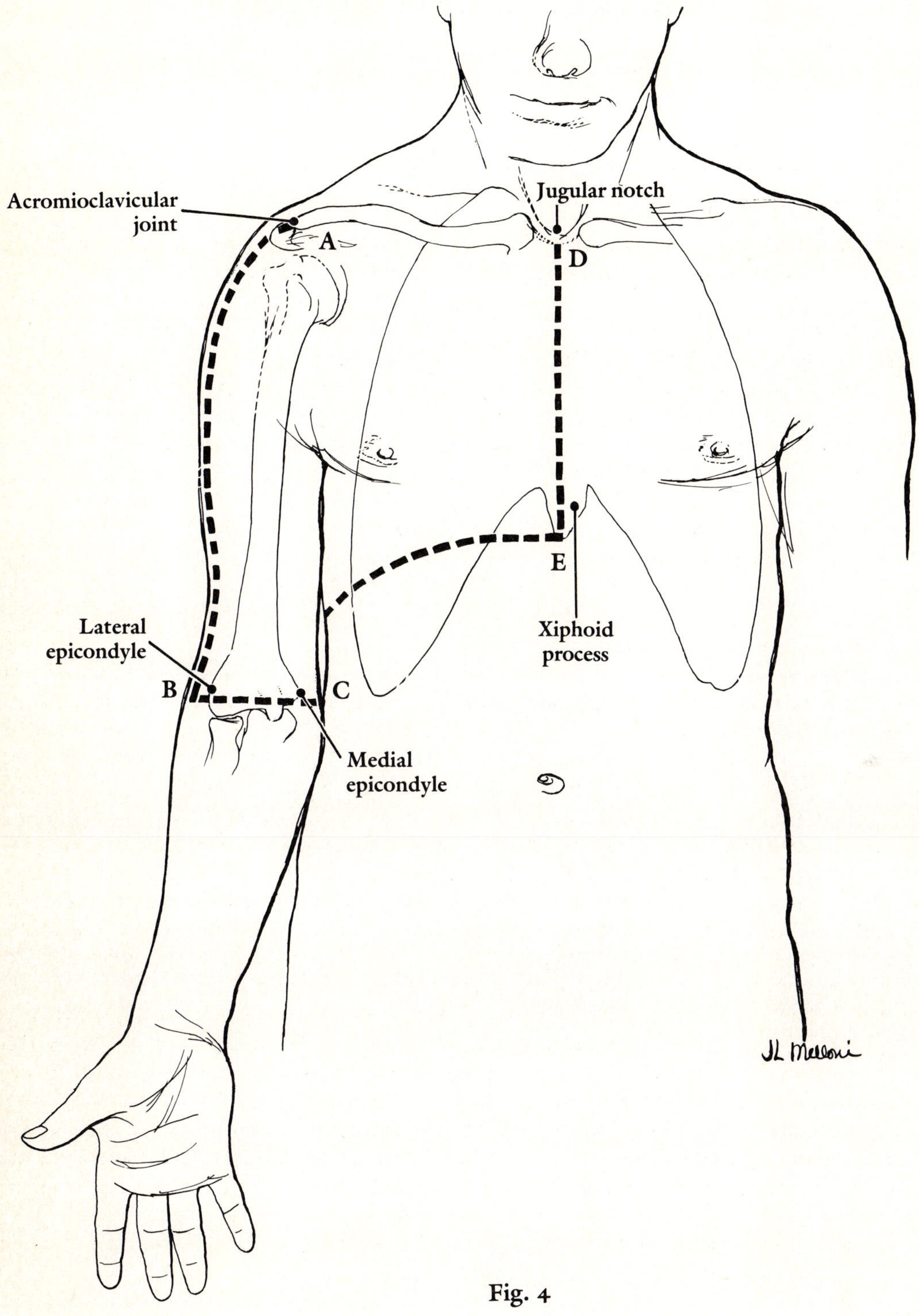
Acromioclavicular joint
A
Jugular notch
D
E
Xiphoid process
Lateral epicondyle
B
C
Medial epicondyle
JL Melloni

Fig. 4

By reflecting the contents of the axilla either laterally or medially, expose on the posterior axillary wall the subscapularis muscle (p. 190). Note the branches of the brachial plexus, the upper and lower subscapular nerves, and the thoracodorsal nerve as they penetrate into the subscapularis and the latissimus dorsi muscles, respectively.

In the course of this dissection, make an attempt to bring into evidence various groups of the axillary lymph nodes.

Summary of structures

1. Arteries
   - Anterior circumflex humeral
   - Axillary
   - Lateral thoracic
   - Pectoral
   - Posterior circumflex humeral
   - Subscapular
   - Superior thoracic
   - Thoracoacromial
2. Bones and bony landmarks
   - Acromion
   - Body of sternum
   - Clavicle
   - Coracoid process
   - Greater tubercle of humerus
   - Jugular notch
   - Lateral epicondyle of humerus
   - Manubrium sterni
   - Medial epicondyle of humerus
   - Ribs
   - Spine of scapula
   - Sternal angle
   - Sternum
   - Xiphoid process
3. Fascia
   - Clavipectoral
4. Glands
   - Areola
   - Lactiferous duct
   - Lobes
   - Mammary
   - Nipple
   - Septa

Table 15

Name	Origin	Insertion	Innervation	Action
Deltoid	Lateral third of clavicle, acromion, and spine of scapula	Deltoid tuberosity of humerus	Axillary nerve	Abducts, flexes, medially rotates, extends, and laterally rotates arm
Latissimus dorsi	Spines of T7 to T12 thoracolumbar fascia (lumbar and sacral spines and posterior part of iliac crest), three lower ribs, and inferior angle of scapula	Intertubercular sulcus of humerus	Thoracodorsal nerve	Adducts, extends, and medially rotates arm
Pectoralis major	Medial half of clavicle, sternum, cartilages of true ribs, and aponeurosis of external oblique muscle of abdomen	Lateral lip of intertubercular sulcus of humerus	Medial and lateral pectoral nerves	Adducts and medially rotates arm
Pectoralis minor	Third, fourth, and fifth ribs	Coracoid process of scapula	Medial and lateral pectoral nerves	Rotates scapula; draws scapula forward
Serratus anterior	Upper eight ribs	Costal surface of medial border of scapula	Long thoracic nerve	Stabilizes and rotates scapula; draws scapula forward
Subclavius	First rib and costoclavicular ligament	Lower aspect of clavicle	Ventral rami of C5 and C6 nerves	Depresses shoulder; draws shoulder forward
Subscapularis	Medial two thirds of subscapular fossa	Lesser tubercle of humerus	Upper and lower subscapular nerves	Stabilizes shoulder; medially rotates arm

5. Joints
   Acromioclavicular
   Sternoclavicular
6. Lymph nodes
   Axillary
7. Muscles
   Deltoid
   Lattisimus dorsi
   Pectoralis major
   Pectoralis minor
   Serratus anterior
   Subclavius
   Subscapularis
8. Nerves
   Axillary
   Intercostobrachial
   Lateral pectoral
   Long thoracic
   Lower subscapular
   Medial antebrachial cutaneous
   Medial brachial cutaneous
   Medial pectoral
   Median
   Musculocutaneous
   Radial
   Thoracodorsal
   Ulnar
   Upper subscapular
9. Plexus
   Brachial
   Brachial, lateral cord
   Brachial, medial cord
   Brachial, posterior cord
10. Regions
   Arm
   Axilla
   Axilla, walls
   Thorax
11. Veins
   Axillary
   Axillary tributaries
   Cephalic
   Pectoral

ANTERIOR REGION OF THE ARM

Dissection. In the lower medial aspect of the arm find the basilic vein coming through the subcutaneous layer from the forearm (p. 182). Follow the vein up to the point where it pierces the deep fascia. Finish the dissection of the medial brachial cutaneous nerve. Incise the deep fascia along the arm, following the anterior midline, and reflect the parts laterally and medially.

Dissect the biceps brachii muscle and observe the two heads in the upper part of the arm (p. 184). Follow the short head, which is medial, up to the coracoid process and expose the long head up to the point where it becomes continuous with the tendon. Immediately medial to the short head, from the coracoid process down to the humerus, dissect the coracobrachialis muscle. This muscle is frequently fused with the short head of the biceps brachii muscle. Note how the musculocutaneous nerve passes through the coracobrachialis muscle.

Dissect (p. 186), from the axilla along the arm, the medial antebrachial cutaneous, median, and ulnar nerves. The ulnar nerve should be traced until it pierces the medial intermus-

Table 16

Name	Origin	Insertion	Innervation	Action
Biceps brachii	Coracoid process and supraglenoid tubercle of scapula	Radial tuberosity	Musculocutaneous nerve	Supinates and flexes forearm; flexes arm
Brachialis	Lower, anterior half of humeral shaft and intermuscular septa	Coronoid process of ulna	Musculocutaneous nerve	Flexes forearm
Coracobrachialis	Coracoid process of scapula	Medial aspect of humeral shaft	Musculocutaneous nerve	Flexes and adducts arm
Lattissimus dorsi (see Table 15)				
Pectoralis major (see Table 15)				
Subscapularis (see Table 15)				

cular septum to enter the posterior region of the arm.

Continue the dissection of the axillary and then the brachial arteries along the medial aspect of the arm and expose the following branches of the latter artery: the profunda brachii, which accompanies the radial nerve, and the superior and inferior ulnar collateral arteries, which originate in the lower part of the arm. Dissect the brachial veins and the tributaries simultaneously.

Cut the biceps brachii muscle horizontally below the junction of the two heads (p. 188) and reflect the lower muscular part onto the forearm.

Continue the dissection of the musculocutaneous nerve after it has passed through the coracobrachialis muscle in the arm and expose its muscular branches.

Dissect the anterior aspect of the brachialis muscle and observe the attachments to the humerus and to both the lateral and the medial septa.

Observe once again the humeral attachments (p. 190) of the pectoralis major, lattissimus dorsi, and subscapularis muscles.

Summary of structures

1. Arteries
   Axillary
   Brachial
   Inferior ulnar collateral
   Profunda brachii
   Superior ulnar collateral
2. Bones and bony landmarks
   Coracoid process
   Humerus
3. Fasciae
   Deep
   Lateral intermuscular septum
   Medial intermuscular septum
4. Muscles
   Biceps brachii
   Biceps brachii, long head
   Biceps brachii, short head
   Brachialis
   Coracobrachialis
   Latissimus dorsi
   Pectoralis major
   Subscapularis
5. Nerves
   Medial antebrachial cutaneous
   Medial brachial cutaneous
   Median
   Musculocutaneous
   Musculocutaneous, muscular branches
   Radial
   Ulnar
6. Regions
   Arm, anterior
   Arm, posterior
   Axilla
   Forearm
7. Veins
   Basilic
   Brachial
   Brachial tributaries

ANTERIOR REGION OF THE FOREARM

Observation. Observe (p. 192) through the skin the distribution of the superficial veins of the forearm (if accessible). By flexing and extending the hand, observe the tendons on the frontal aspect of the wrist.

Palpation. Palpate the muscular masses arising from the medial and lateral epicondyles of the humerus and between them the cubital fossa with the tendon of the biceps brachii muscle in its floor. Identify the tendons of the muscles arising from the medial epicondyle at the wrist. Palpate, through the soft tissues, the distal extremity of the humerus with the medial and lateral epicondyles, the proximal extremities of the ulna and the radius (p. 176), shafts of both bones, and their distal extremities and the styloid processes (p. 178).

Dissection. Skin incision (Fig. 5): (1) the upper horizontal incision has already been made; (2) make the lower horizontal incision at the level of and from the styloid process of the radius medially to the styloid process of the ulna; (3) make a vertical cut from the upper to the lower horizontal incisions, along the anterior midline of the forearm.

Reflect the skin laterally and medially.

In the subcutaneous tissue (p. 194), follow the cephalic and basilic veins from the arm down and dissect their main communication, the median vein of the forearm.

Complete the dissection of the medial antebrachial cutaneous and musculocutaneous (lateral antebrachial cutaneous) nerves from the arm throughout the forearm.

Expose the tendon of the biceps brachii muscle as far as its radial attachment, and from this tendon observe the bicipital aponeurosis, which extends medially to the deep fascia. Cut the aponeurosis close to and along the bicipital tendon.

Cut the antebrachial fascia vertically, following the anterior midline of the forearm, and reflect the parts medially and laterally (p. 196).

Dissect, from the lateral side of the humerus to the styloid process of the radius the brachioradialis muscle. Reflect the muscle slightly to the lateral side, and on its deep aspect

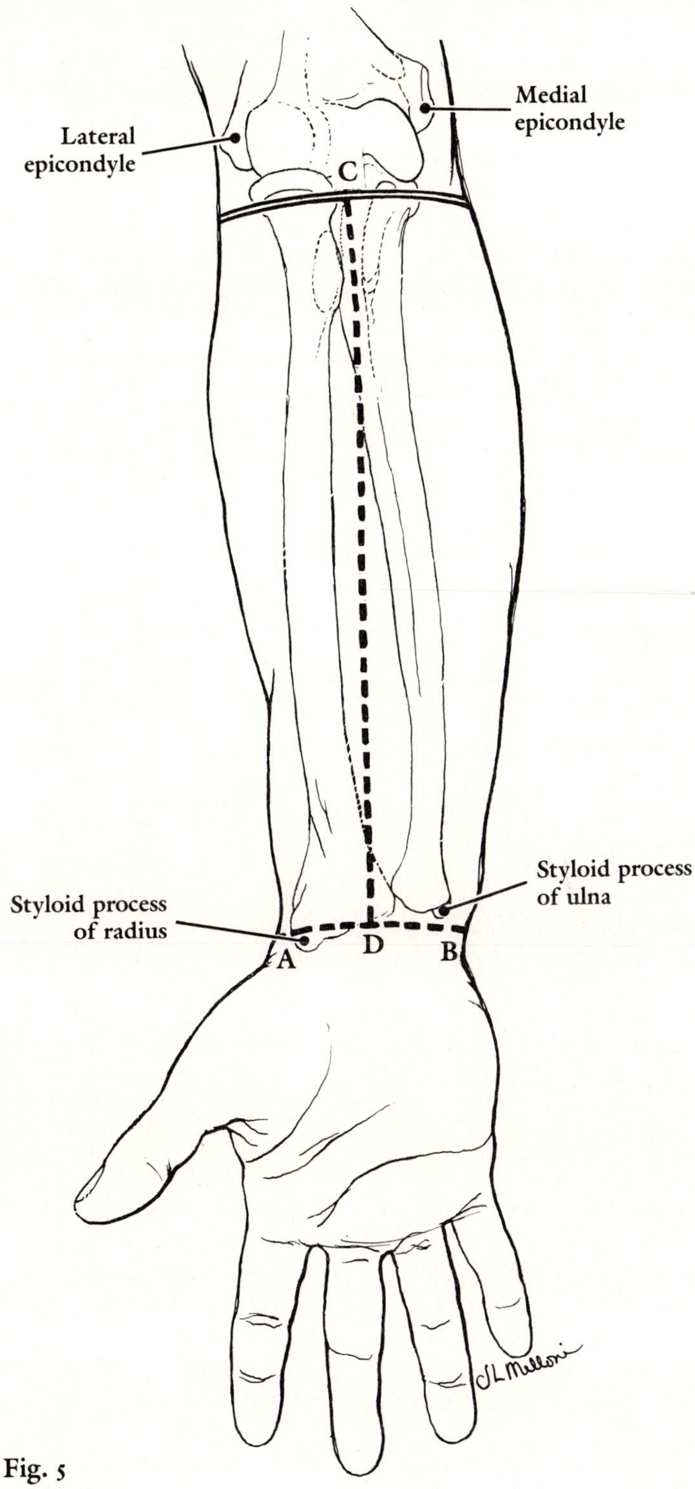
Medial
epicondyle
Lateral
epicondyle
C
Styloid process
of ulna
Styloid process
of radius
A
D
B

Fig. 5

find the superficial branch of the radial nerve. Follow the branch until it pierces the fascia behind the tendon of the brachioradialis muscle and up to the bifurcation of the radial nerve, in the intermuscular space between the brachialis and the brachioradialis muscles. Identify the origin of the deep branch from the bifurcation of the radial neve. Continue the dissection of the brachial artery down to the bifurcation, and from there through the forearm expose the radial artery. Dissect the radial recurrent branch of this artery in the upper part of the forearm. Expose the initial part of the ulnar artery, also starting from the bifurcation.

Follow the median nerve down as it enters into the musculature of the medial epicondyle.

Dissect, from the medial epicondyle, the anterior aspects of (1) the pronator teres muscle (the superficial head) to the radius, (2) the flexor carpi radialis muscle to the wrist, (3) the palmaris longus muscle, if present, to the wrist, and (4) the flexor carpi ulnaris muscle to the wrist.

Table 17

Name	Origin	Insertion	Innervation	Action
Biceps brachii (see Table 16)				
Brachialis (see Table 16)				
Brachioradialis	Lateral supracondylar ridge of humerus and lateral intermuscular septum	Radius, above styloid process	Radial nerve	Flexes arm
Flexor carpi radialis	Medial epicondyle of humerus, intermuscular septum, and antebrachial fascia	Base of second metacarpal bone	Median nerve	Flexes hand
Flexor carpi ulnaris	Medial epicondyle of humerus and medial margin of olecranon	Pisiform bone	Ulnar nerve	Flexes hand
Flexor digitorum profundus	Anterior and posterior aspects of ulna, coronoid process, and interosseous membrane	Base of distal phalanges	Median and ulnar nerves	Flexes distal phalanges and hand
Flexor digitorum superficialis	Medial epicondyle of humerus, elbow joint, intermuscular septum, coronoid process, and anterior border of radial shaft	Shaft of middle phalanges (divided tendons)	Median nerves	Flexes middle and proximal phalanges and wrist
Flexor pollicis longus	Anterior aspect of radial shaft and interosseous membrane	Base of distal phalanx of thumb	Median nerves	Flexes phalanges of thumb
Palmaris longus	Medial condyle of humerus, intermuscular septum, and antebrachial fascia	Flexor retinaculum and palmar aponeurosis	Median nerves	Flexes hand; tensor of palmar fascia
Pronator quadratus	Lower part of anterior surface of ulnar shaft	Lower part of anterior surface of radial shaft	Median nerve	Pronates forearm
Pronator teres	Medial epicondyle of humerus, intermuscular septum, antebrachial fascia, and coronoid process of ulna	Radial shaft	Median nerve	Pronates and flexes forearm

In between the tendons of the brachioradialis and the flexor carpi radialis muscles find the radial artery; in between the tendons of the flexor carpi radialis and palmaris longus muscles and at a deeper plane find the median nerve; and in between the tendons of the palmaris longus and the flexor carpi ulnaris muscles find the ulnar nerve and blood vessels.

Cut horizontally through the middle of the flexor carpi radialis and the palmaris longus muscles and reflect the parts to their respective attachments.

Dissect the anterior aspect of the flexor digitorum superficialis muscle (p. 198) and observe its four tendons. Remove this muscle from the radius, cut it horizontally through the middle, and taking care not to sever the median nerve, which clings to its deep surface, reflect the parts to their respective attachments. Observe (p. 200) how the median nerve passes between the two heads, the superficial and the deep, of the pronator teres muscle. Dissect the nerve down to the wrist. Expose its branches for the muscles in the anterior region of the forearm. Cut vertically through the superficial head of the pronator teres muscle and reflect the parts to their respective attachments. Expose the deep head.

Follow the ulnar artery from its origin as it passes deep to the deep head of the pronator teres muscle.

Dissect the anterior aspect of the flexor digitorum profundus and the flexor pollicis longus muscles.

Cut the flexor digitorum profundus muscle horizontally through the middle and reflect the parts to their respective attachments (p. 202).

Cut the deep head of the pronator teres muscle vertically and complete the dissection of the ulnar artery in the forearm. Expose its branches in the upper part of the forearm—the ulnar recurrent and the common interosseous artery. The latter branch bifurcates into the anterior and posterior interosseous arteries. Follow the anterior interosseous artery, which is accompanied by the interosseous branch of the median nerve, down to the pronator quadratus muscle. Follow the posterior interosseous artery until it passes through the interosseous membrane.

Complete the dissection of the ulnar nerve up to where it passes behind the medial epicondyle of the humerus.

Dissect the anterior aspect of the pronator quadratus muscle. Clean the interosseous membrane.

Summary of structures

1. Arteries
    - Anterior interosseous
    - Brachial
    - Common interosseous
    - Posterior interosseous
    - Radial
    - Radial recurrent
    - Ulnar
    - Ulnar recurrent
2. Bones and bony landmarks
    - Humerus
    - Lateral epicondyle
    - Medial epicondyle
    - Radius
    - Styloid process of radius
    - Styloid process of ulna
    - Ulna
3. Fasciae
    - Bicipital aponeurosis
    - Deep
    - Interosseous membrane
4. Muscles
    - Biceps brachii tendon
    - Brachialis
    - Brachioradialis
    - Flexor carpi radialis
    - Flexor carpi ulnaris
    - Flexor digitorum profundus
    - Flexor digitorum superficialis
    - Flexor pollicis longus
    - Palmaris longus
    - Pronator quadratus
    - Pronator teres
    - Pronator teres, deep head
    - Pronator teres, superficial head
5. Nerves
    - Lateral antebrachial cutaneous
    - Medial antebrachial cutaneous
    - Median
    - Median, interosseous branch
    - Median, muscular branches
    - Musculocutaneous
    - Radial
    - Radial, deep branch
    - Radial, superficial branch
    - Ulnar
6. Regions
    - Cubital fossa
    - Forearm
7. Veins
    - Basilic
    - Cephalic
    - Median of the forearm
    - Superficial
    - Ulnar

ANTERIOR REGION OF THE HAND

Observation. Observe, on the anterior surface of the hand (p. 192), the thenar and hypothenar eminences and between them the middle zone of the hand. Study the main sulci.

Palpation. Palpate (p. 178) the pisiform bone and the hook of the hamate bone, the scaphoid and trapezium bones, the lateral and medial edges of the hand skeleton, the metacarpal bones, the phalanges, and the thenar and hypothenar musculatures.

Dissection. Skin incision (Fig. 6): (1) cut along the medial edge of the hand to the root of the little finger; (2) cut along the lateral edge of the hand to the root of the thumb, then across the anterior aspect of the root of the thumb, and proceed to the root of the index finger; (3) the upper horizontal incision has already been made; (4) make the lower horizontal incision from the root of the index finger medially across the roots of the third and fourth fingers to the root of the little finger; (5) incise the skin of each digit longitudinally down its midline.

Reflect the flaps of the skin from the digits medially and laterally. Remove the skin from the anterior aspect of the hand.

Dissect in the hypothenar subcutaneous layer (p. 194) and subsequently remove the palmaris brevis muscle. This muscle is frequently absent or poorly developed. Clean the hypothenar fascia.

At the continuation of the tendon of the palmaris longus muscle to the roots of the last four digits, expose the palmar aponeurosis and its four digital bands. Note how the bands are kept together by the superficial transverse fibers.

Clean the surface of the thenar fascia.

Remove the palmar aponeurosis and the four bands by cutting through the medial and lateral septa along the

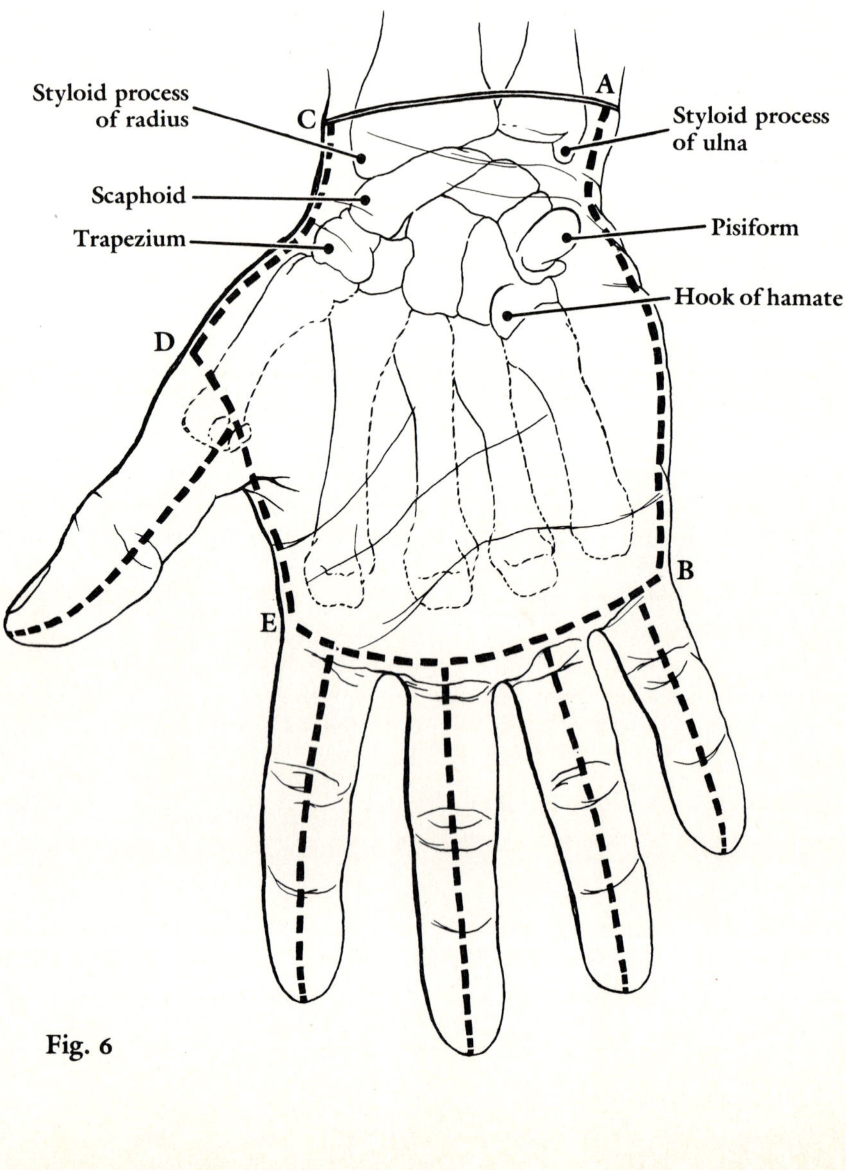

Fig. 6

medial and lateral borders of the aponeurosis, respectively (p. 204).

Dissect the superficial palmar arch (located under the palmar aponeurosis) medially to the ulnar artery and laterally to the radial artery. The connection of the arch to the radial artery is often absent. Dissect the digital branches of the superficial palmar arch to the respective fingers.

Clean the anterior aspect of the flexor retinaculum and expose precisely its inferior border.

Dissect the median nerve in the hand and identify the origins of its digital (for three and one-half digits laterally), recurrent, and muscular branches.

Dissect the ulnar nerve in the hand until it bifurcates and then the superficial and deep branches. Follow, from the superficial branch, the three digital branches for the medial half of the fourth and the medial and lateral halves of the fifth digits.

Throughout the middle zone of the hand and along the fingers clean the synovial sheaths and their ligaments (p. 206). These sheaths contain the tendons of the flexor digitorum superficialis and profundus muscles. Cut the successive sheaths longitudinally and expose the tendons of the flexor digitorum superficialis muscle as far as their bifurcations and their attachments at the bases of the middle phalanges.

Remove the thenar fascia and dissect the abductor pollicis brevis muscle. Cut the muscle horizontally through its middle and reflect the parts to their respective attachments.

Table 18

Name	Origin	Insertion	Innervation	Action
Abductor digiti minimi	Pisiform bone and pisohamate ligament	Proximal phalanx of little finger	Deep branch of ulnar nerve	Abducts little finger
Abductor pollicis brevis	Flexor retinaculum, scaphoid and trapezium bones	Base of proximal phalanx of thumb	Median nerve	Abducts and medially rotates thumb
Adductor pollicis	Capitate and second and third metacarpal bones	Base of proximal phalanx of thumb	Deep branch of ulnar nerve	Adducts thumb
Flexor digiti minimi brevis	Hook of hamate bone and flexor retinaculum	Base of proximal phalanx of little finger	Deep branch of ulnar nerve	Flexes little finger
Flexor digitorum profundus (see Table 17)				
Flexor digitorum superficialis (see Table 17)				
Flexor pollicis brevis	Flexor retinaculum, trapezium, trapezoid, and capitate bones	Base of proximal phalanx of thumb	Median nerve and deep branch of ulnar nerve	Flexes proximal phalanx of thumb
Flexor pollicis longus (see Table 17)				
Lumbricals	Tendons of flexor digitorum profundus	Radial side of corresponding finger and dorsal digital expansion	Median nerve and deep branch of ulnar nerve	Flex proximal phalanges; extend middle and distal phalanges
Opponens digiti minimi	Hook of hamate bone and flexor retinaculum	Ulnar margin of fifth metacarpal bone	Deep branch of ulnar nerve	Flexes and laterally rotates fifth metacarpal bone
Opponens pollicis	Trapezium bone and flexor retinaculum	Lateral border of first metacarpal bone	Median nerve	Flexes and medially rotates first metacarpal bone
Palmar interossei	Palmar surfaces of metacarpal bones	Base of dorsal digital expansion	Deep branch of ulnar nerve	Adducts fingers; flexes proximal phalanges; extends middle and distal phalanges
Palmaris brevis	Flexor retinaculum and palmar aponeurosis	Dermis of ulnar side of hand	Superficial branch of ulnar nerve	Wrinkles skin over ulnar side of hand
Palmaris longus (see Table 17)				

Expose in the next layer the flexor pollicis brevis muscle, the tendon of the flexor pollicis longus muscle, and the opponens pollicis muscle.

Dissect in the hypothenar eminence the abductor digiti minimi and the flexor digiti minimi brevis muscles. Reflect the tendons of the flexor digitorum superficialis muscle to their respective attachments by drawing them forward from the carpal canal (tunnel), behind the flexor retinaculum.

Dissect the tendons of the flexor digitorum profundus muscle together with the lumbrical muscles through the hand and fingers.

Make two vertical, parallel incisions through the middle of the flexor retinaculum, 1 cm apart, and remove the middle part. Reflect the tendons of the flexor digitorum profundus muscle, together with the associated lumbrical muscles, from the carpal canal and the hand onto the digits. Remove the digital branches of the superficial palmar arch and the median and ulnar nerves.

Remove the flexor digiti minimi brevis and the abductor digiti minimi muscles in the hypothenar eminence and the flexor pollicis brevis and the opponens pollicis muscles in the thenar eminence (p. 208).

Dissect in the thenar eminence the adductor pollicis muscle. In between the two heads of this muscle identify the deep palmar arch, which is the continuation of the radial artery.

Dissect the opponens digiti minimi muscle in the hypothenar eminence and subsequently remove it.

Follow the deep branch of the ulnar nerve while exposing the palmar interosseous muscles in their respective intermetacarpal spaces. Verify how these muscles relate to the corresponding digits.

Summary of structures

1. Arteries
   Deep palmar arch
   Digital branches of superficial palmar arch
   Radial
   Superficial palmar arch
   Ulnar
2. Bones and bony landmarks
   Hamate
   Hand skeleton
   Hook of hamate
   Intermetacarpal spaces
   Metacarpals
   Phalanges
   Pisiform
   Scaphoid
   Trapezium
3. Fasciae
   Flexor retinaculum
   Hypothenar
   Lateral septum
   Medial septum
   Palmar aponeurosis
   Palmar aponeurosis, digital bands
   Superficial transverse fibers
   Synovial sheaths
   Synovial sheaths, ligaments
   Thenar
4. Muscles
   Abductor digiti minimi
   Abductor pollicus brevis
   Adductor pollicis
   Flexor digiti minimi brevis
   Flexor digitorum profundus
   Flexor digitorum superficialis
   Flexor pollicis brevis
   Flexor pollicis longus
   Hypothenar
   Lumbricals
   Opponens digiti minimi
   Opponens pollicis
   Palmar interossei
   Palmaris brevis
   Palmaris longus
   Thenar
5. Nerves
   Digital branches of median
   Digital branches of ulnar
   Median
   Median, muscular branches
   Median, recurrent branch
   Ulnar
   Ulnar, deep branch
   Ulnar, superficial branch
6. Surface anatomy
   Fourth finger
   Hand
   Hypothenar eminence
   Index finger
   Little finger
   Middle zone of hand
   Palmar sulci
   Thenar eminence
   Third finger
   Thumb
7. Tunnel
   Carpal

SCAPULAR AND DELTOID REGIONS AND THE POSTERIOR REGION OF THE ARM

Preliminary dissection. Remove all muscles from the medial half of the clavicle and the sternum. Dissect the sternoclavicular joint, the anterior aspect of the articular capsule, the anterior sternoclavicular ligament, the interclavicular ligament, and the costoclavicular ligament. Remove the anterior half of the articular capsule and observe, deep to it, the articular surface and the intraarticular disc. Cut the remaining part of the articular capsule vertically and the subclavius muscle through its middle, and reflect the clavicle laterally. Complete the dissection of the axillary artery and vein and the brachial plexus from the axillary fossa into the neck.

Expose, by rotating the upper limb as a whole laterally, the serratus anterior muscle as far as its insertion into the vertebral, medial margin of the scapula.

Observe once again, behind the contents of the axillary fossa, the entire anterior surface of the subscapularis muscle.

Turn the cadaver face downward.

Observation. Observe the olecranon, deltoid muscle, latissimus dorsi muscle, and triceps brachii muscle (pp. 57 and 173).

Palpation. Palpate the observed structures (p. 169).

Dissection. By a vertical incision (Fig. 7) separate the skin of the posterior aspect of the arm from that of the

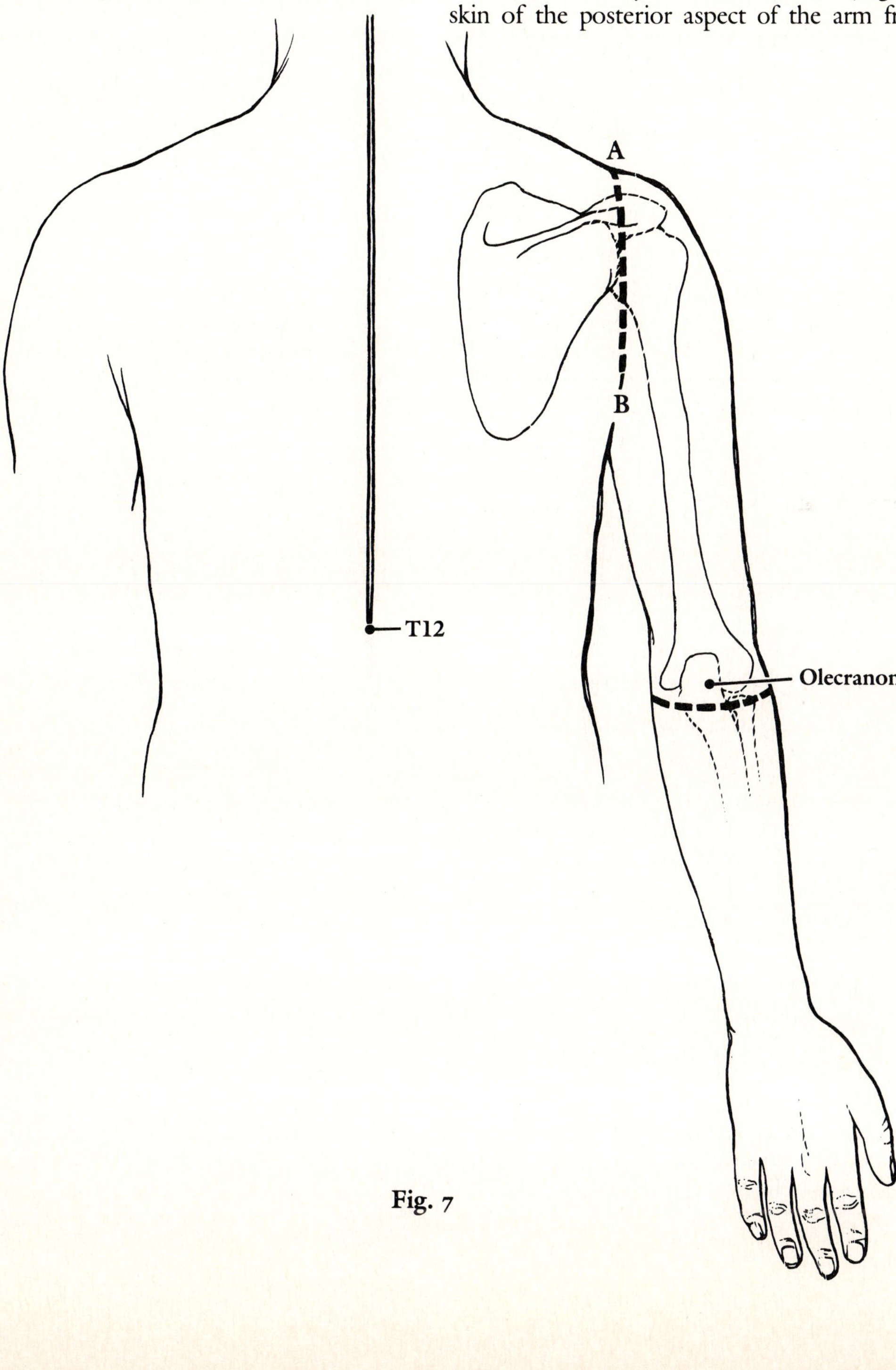

Fig. 7

scapular region and reflect the skin flaps to the olecranon and the spine of the T12 vertebra, respectively.

Review the attachments of the trapezius muscle, rhomboideus muscles, and the levator scapulae muscle to the scapula (pp. 58 and 60).

Complete the dissection of the deltoid muscle (p. 210) from the acromion and the spine of the scapula down to the humerus. Remove the muscle from the clavicle, acromion, and the spine of the scapula and reflect it halfway toward its humeral attachment (p. 212). Observe the subdeltoid bursa on the deep aspect of the muscle and the branches of the axillary nerve that innervate the muscle.

Dissect and subsequently free that part of the latissimus dorsi muscle which extends from the posterior surface of the thorax to the humerus. Separate precisely this from the other scapular muscles.

Dissect from the scapula to the greater tubercle of the humerus, in superoinferior order, the supraspinatus, infraspinatus, and the teres minor muscles.

Remove the fascia from the posterior aspect of the arm and expose the long and lateral heads of the triceps brachii muscle. Note the relationship of the teres minor muscle to the long head of the triceps brachii muscle. Dissect from the inferior angle of the scapula to the humerus the teres major muscle and appreciate its relationship with the long head of the triceps brachii muscle.

Explore the triangular space between the teres minor muscle superiorly, the teres major muscle inferiorly, and the long head of the triceps brachii muscle laterally. Within the triangular space find the circumflex scapular artery and vein.

Explore the quadrangular space between the teres minor muscle superiorly, the teres major muscle inferiorly, the long head of the triceps brachii muscle medially, and the humerus laterally. Expose within the quadrangular space the axillary nerve and the posterior circumflex humeral artery and vein. Complete the dissection of the axillary nerve to the deltoid muscle and reflect the muscle all the way to the humerus.

Explore another triangular space between the teres major muscle superiorly, the long head of the triceps brachii muscle medially, and the lateral head of the triceps brachii muscle laterally. Dissect within this space the radial nerve and the profunda brachii vessels.

By reflecting partially the supraspinatus muscle, expose

Table 19

Name	Origin	Insertion	Innervation	Action
Infraspinatus	Medial two thirds of infraspinous fossa of scapula and infraspinous fascia	Greater tubercle of humerus	Suprascapular nerve	Stabilizes shoulder; abducts arm
Latissimus dorsi (see Table 15)				
Levator scapulae (see Table 7)				
Rhomboideus (see Table 8)				
Serratus anterior (see Table 15)				
Subclavius (see Table 15)				
Subscapularis (see Table 15)				
Supraspinatus	Medial two thirds of supraspinous fossa of scapula and supraspinous fascia	Greater tubercle of humerus	Suprascapular nerve	Stabilizes shoulder; abducts arm
Teres major	Dorsal aspect of inferior angle of scapula	Medial lip of intertubercular sulcus of humerus	Lower subscapular nerve	Adducts, extends, and medially rotates arm
Teres minor	Upper two thirds of lateral border of scapula	Greater tubercle of humerus	Axillary nerve	Stabilizer of shoulder; abducts and laterally rotates arm
Trapezius (see Table 5)				
Triceps brachii	Infraglenoid tubercle of scapula and posterior aspect of humeral shaft	Olecranon of ulna and antebrachial fascia	Radial nerve	Extends forearm; adducts and extends arm

along the upper border of the scapula the suprascapular ligament and the suprascapular blood vessels and nerve (p. 214).

Reflect the levator scapulae and rhomboideus muscles onto the scapula, and deep to them complete the dissection of the deep branch of the transverse cervical artery and the dorsal scapular nerve.

Explore the intermuscular space between the long and lateral heads of the triceps brachii muscle down toward the olecranon by separating them from one another. Reflect the long head medially.

Follow the radial nerve and the profunda brachii vessels through the arm. Isolate the branches of the nerve to the heads of the triceps brachii muscle and the radial collateral branch of the profunda brachii artery.

Finish the dissection of the medial head of the triceps brachii muscle to the olecranon.

Separation of the upper limb from the trunk. Cut vertically through the free part of the latissimus dorsi muscle. Reflect the vertebral border of the scapula laterally and posteriorly. Demonstrate the scapular attachment of the serratus anterior muscle and cut the muscle vertically about 2 cm lateral to the vertebral border of the scapula.

Cut the brachial plexus and the axillary blood vessels horizontally at the level of the clavicle. Separate the upper limb from the trunk.

Summary of structures

1. Arteries
   Axillary
   Circumflex scapular
   Posterior circumflex humeral
   Profunda brachii
   Radial collateral
   Suprascapular
   Transverse cervical, deep branch
2. Bones and bony landmarks
   Acromion
   Clavicle
   Greater tubercle of humerus
   Humerus
   Inferior angle of scapula
   Medial border of scapula
   Olecranon
   Scapula
   Spine of scapula
   Sternum
   Superior border of scapula
   T12 spine
3. Bursa
   Subdeltoid
4. Joints
   Articular capsule
   Articular surfaces
   Intraarticular disc
   Sternoclavicular
5. Ligaments
   Anterior sternoclavicular
   Costoclavicular
   Interclavicular
   Suprascapular
6. Muscles
   Deltoid
   Infraspinatus
   Latissimus dorsi
   Levator scapulae
   Rhomboideus
   Scapular
   Serratus anterior
   Subclavius
   Subscapularis
   Supraspinatus
   Teres major
   Teres minor
   Trapezius
   Triceps brachii
   Triceps brachii, lateral head
   Triceps brachii, long head
   Triceps brachii, medial head
7. Nerves
   Axillary
   Axillary branches
   Dorsal scapular
   Radial
   Radial, muscular branches
   Suprascapular
8. Plexus
   Brachial
9. Regions
   Arm, posterior
   Axilla
   Neck
   Scapular
   Thorax
   Trunk
   Upper limb
10. Spaces
    Quadrangular, scapular
    Triangular, humeral
    Triangular, scapular
11. Veins
    Axillary
    Circumflex scapular
    Posterior circumflex humeral
    Profunda brachii
    Suprascapular

POSTERIOR REGIONS OF THE FOREARM AND THE HAND

Observation. Observe (p. 193), if accessible, the subcutaneous veins and the position of the styloid processes of the ulna and the radius through the skin of the forearm and the hand.

Palpation. Palpate the olecranon, the posterior border of the ulna (p. 177), the skeletal components of the wrist joint and the hand (p. 179), and the tendons on the posterior aspect of the wrist and the hand.

Dissection. Remove the skin (Fig. 8) from the posterior aspect of the forearm, hand, and fingers.

Dissect, in the subcutaneous tissue (p. 216), the venous arch on the dorsum of the hand and from it the initial parts of the basilic and cephalic veins. Complete the dissection of the superficial branch of the radial nerve and of the dorsal branch of the ulnar nerve in the hand. Study the distribution of these branches to the skin of the posterior aspect of the hand and fingers.

Remove the fascia from the forearm and define precisely the posterior aspect and both the upper and the lower borders of the extensor retinaculum (p. 218).

Dissect from the lateral humeral epicondyle to the upper border of the extensor retinaculum, from lateral to medial, the following muscles: the extensor carpi radialis longus, the extensor carpi radialis brevis, the extensor digitorum, the extensor digiti minimi, the extensor carpi ulnaris, and the anconeus (this extends to the posterior border of the ulna).

Complete the dissection of the humeral and ulnar heads of the flexor carpi ulnaris muscle on the medial aspect of the ulna and, in between the two heads, the ulnar nerve.

Note how the abductor pollicis longus and the extensor pollicis brevis muscles cross the posterior aspect of the tendons of the two extensor carpi radialis muscles. The two former muscles arise from the deep muscular layer, cross the posterior aspect of the extensor carpi radialis muscles, and then descend on the lateral side of the wrist. Expose these two muscles up to the upper margin of the extensor retinaculum.

Cut horizontally through the middle of the extensor digitorum, extensor digiti minimi, and the extensor carpi ulnaris muscles, and reflect their parts superiorly and inferiorly.

In the deep muscular layer (p. 220) dissect the supinator muscle and the deep branch of the radial nerve as it passes through the muscle. In addition to this muscle, within the same muscular layer, complete the dissection of the abductor pollicis longus and the extensor pollicis brevis muscles. Dissect the extensor pollicis longus and the extensor indicis muscles down, toward the upper border of the extensor retinaculum.

Identify the tendons of the posterior antebrachial muscles on the dorsum of the hand. Dissect further these tendons from the lower border of the extensor retinaculum to

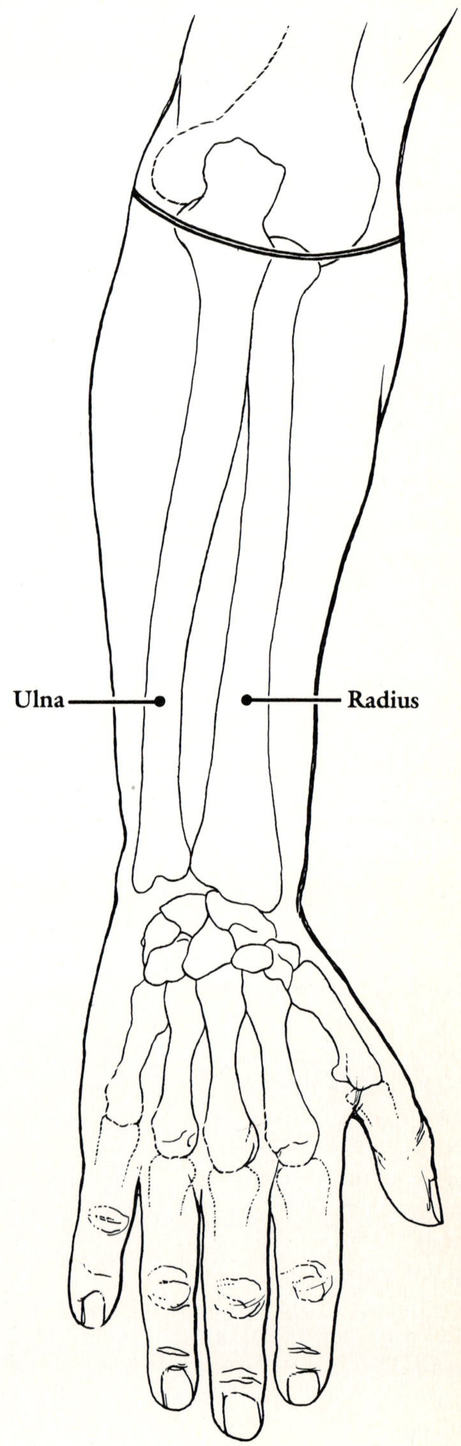

Fig. 8

Table 20

Name	Origin	Insertion	Innervation	Action
Abductor pollicis longus	Posterior surface of ulnar shaft, interosseous membrane, and radius	Radial side of first metacarpal and trapezium bones	Posterior interosseous nerve	Abducts and extends thumb
Anconeus	Posterior surface of lateral epicondyle of humerus	Lateral side of olecranon and upper fourth of ulnar shaft	Radial nerve	Extends and pronates forearm
Dorsal interossei	Adjacent metacarpal bones	Base of proximal phalanges and dorsal digital expansions	Deep branch of ulnar nerve	Abduct fingers; flex proximal phalanges; extend middle and distal phalanges
Extensor carpi radialis brevis	Lateral epicondyle of humerus, radial collateral ligament, and intermuscular septa	Dorsal surface of base of third metacarpal bone	Posterior interosseous nerve	Extends and abducts hand
Extensor carpi radialis longus	Lateral supracondylar ridge of humerus and lateral intermuscular septum	Dorsal surface of base of second metacarpal bone	Radial nerve	Extends and abducts hand
Extensor carpi ulnaris	Lateral epicondyle of humerus, posterior border of ulna, and antebrachial fascia	Medial side of base of fifth metacarpal bone	Posterior interosseous nerve	Extends and adducts hand
Extensor digiti minimi	Common extensor tendon and intermuscular septa	Dorsal digital expansion of fifth digit	Posterior interosseous nerve	Extends hand and little finger
Extensor digitorum	Lateral epicondyle of humerus, intermuscular septa, and antebrachial fascia	Dorsal digital expansion	Posterior interosseous nerve	Extends hand and fingers
Extensor indicis	Posterior surface of ulna and interosseous membrane	Dorsal digital expansion	Posterior interosseous nerve	Extends index finger and hand
Extensor pollicis brevis	Posterior surface of radius and interosseous membrane	Base of proximal phalanx of thumb	Posterior interosseous nerve	Extends thumb and first metacarpal bone
Extensor pollicis longus	Posterior surface of ulna and interosseous membrane	Base of distal phalanx of thumb	Posterior interosseous nerve	Extends distal and proximal phalanges of thumb and first metacarpal bone
Flexor carpi ulnaris (see Table 17)				
Lumbricals (see Table 18)				
Supinator	Lateral epicondyle of humerus, radial collateral ligament, annular ligament, supinator crest of ulna, and supinator aponeurosis	Lateral surface of proximal third of radial shaft	Posterior interosseous nerve	Supinates forearm

their respective attachments in the hand and fingers.

Note the boundaries of the anatomical snuffbox: the abductor pollicis longus and extensor pollicis brevis tendons laterally and the extensor pollicis longus tendon medially. Expose in the anatomical snuffbox the radial artery, and follow the artery as it proceeds through the first intermetacarpal space to the palmar region, where it becomes continuous with the deep palmar arch. Dissect the main collateral branch of the artery, the dorsal metacarpal, which extends toward the medial margin of the hand.

By six vertical incisions through the extensor retinaculum, open the six osteofibrous canals (tunnels) and demonstrate their contents. The following tendons, from lateral to medial, are found in these canals (tunnels): abductor pollicis longus and extensor pollicis brevis, extensor carpi radialis longus and brevis, extensor pollicis longus, extensor digitorum and indicis, extensor digiti minimi, and extensor carpi ulnaris.

While dissecting the tendons of the extensor digitorum muscle into the fingers, note the dorsal expansion to which the interosseous and lumbrical muscles are attached.

Dissect the dorsal interosseous muscles.

Find the posterior interosseous artery in the forearm, where it passes through the interosseous membrane. Follow the artery along the posterior aspect of the interosseous membrane by gradually removing the muscles.

Summary of structures

1. Arteries
   - Deep palmar arch
   - Dorsal metacarpal
   - Posterior interosseous
   - Radial
2. Bones and bony landmarks
   - Hand skeleton
   - Lateral epicondyle of humerus
   - Medial surface of ulna
   - Olecranon
   - Posterior border of ulna
   - Styloid process of radius
   - Styloid process of ulna
   - Wrist skeleton
3. Fasciae
   - Deep, of forearm
   - Dorsal digital expansions
   - Extensor retinaculum
   - Interosseous membrane of forearm
4. Muscles
   - Abductor pollicis longus
   - Anconeus
   - Dorsal interossei
   - Extensor carpi radialis brevis
   - Extensor carpi radialis longus
   - Extensor carpi ulnaris
   - Extensor digiti minimi
   - Extensor digitorum
   - Extensor indicis
   - Extensor pollicis brevis
   - Extensor pollicis longus
   - Flexor carpi ulnaris, humeral head
   - Flexor carpi ulnaris, ulnar head
   - Interossei
   - Lumbricals
   - Supinator
5. Nerves
   - Radial, deep branch
   - Radial, superficial branch
   - Ulnar
   - Ulnar, dorsal branch
6. Surface anatomy
   - Anatomical snuffbox
   - Fingers
   - Forearm
   - Hand
   - Medial margin of hand
   - Palmar region
   - Subcutaneous veins
7. Tunnels
   - Osteofibrous canals
8. Veins
   - Basilic
   - Cephalic
   - Dorsal venous arch

JOINTS OF THE UPPER LIMB

Acromioclavicular joint (pp. 172, 173, 224, and 225). Expose on the superior aspect of the acromioclavicular joint the articular capsule and on the anterior aspect the coracoclavicular and coracoacromial ligaments. Open the joint by cutting the articular capsule circularly, and study the articular surfaces.

Glenohumeral (shoulder) joint (pp. 168 to 170). Remove the subscapularis muscle from the scapula and reflect it toward the humerus. Observe, under cover of the muscle, the subscapular bursa. Separate the bursa from the deep surface of the muscle; preserve it in situ on the anterior aspect of the articular capsule. Remove the teres major muscle. Expose the anterior aspect of the joint (p. 224); note the three glenohumeral ligaments (superior, middle, and inferior) and the opening between the superior and middle ligaments.

Remove the supraspinatus, infraspinatus, and teres

minor muscles from the scapula and reflect them onto the humerus. Dissect on the superior aspect of the joint the coracohumeral ligament (p. 225). Expose the posterior aspect of the articular capsule. Subsequently cut the capsule circularly and reflect the parts to their respective attachments. Observe the attachment and the relationships of the tendon of the long head of the biceps brachii muscle within the joint.

Elbow joint (pp. 176 and 177). Reflect the biceps brachii muscle to the radial attachment and remove the brachialis muscle from the coronoid process of the ulna. Remove gradually all muscles from the lateral and medial epicondyles of the humerus.

Reflect the triceps brachii muscle on the olecranon and observe its bursa, which is interposed between the muscle and the bone. Remove the muscle.

Dissect in the flexed forearm the posterior aspect of the articular capsule (p. 227) and on the medial and lateral sides the ulnar collateral and radial collateral ligaments, respectively.

Dissect in the extended forearm the anterior aspect of the articular capsule (p. 226) and the annular ligament. Cut circularly, above the annular ligament, through the articular capsule and separate the humerus from the radius and ulna. Study the proximal radioulnar joint. Cut vertically through the annular ligament and separate the radius from the ulna.

Wrist joint (pp. 178 and 179). Reflect all muscles from the surfaces of the wrist joint toward their attachments in the hand and fingers.

With the hand extended (p. 228), dissect the anterior aspect of the articular capsule and the following ligaments, from lateral to medial: radial collateral, palmar radiocarpal, palmar ulnocarpal, and ulnar collateral.

Expose on the anterior aspect of the hand the following ligaments: pisohamate, pisometacarpal, palmar intercarpal, carpometacarpal, and intermetacarpal.

With the hand flexed (p. 230), dissect the posterior aspect of the articular capsule and the following ligaments, from lateral to medial: radial collateral, dorsal radiocarpal, ulnar collateral.

Expose the dorsal intercarpal ligaments on the posterior aspect of the hand.

Open the joint (p. 229) by cutting through the articular capsule anteriorly. Remove carefully the capsule from the anterior aspects of the radius, ulna, intraarticular disc, and the carpal bones. Observe the disc, which extends from the radius, between the head of the ulna and the carpal bones medial to the styloid process of the ulna. By cutting vertically, open the distal radioulnar joint.

Carpal joints (pp. 178 and 179). Cut through the ligaments and the articular capsules to separate the carpal bones from one another and from the metacarpals.

Metacarpophalangeal joints (pp. 178 and 179). Open all metacarpophalangeal joints.

Interphalangeal joints (pp. 178 and 179). Open all interphalangeal joints.

NOTE: In the course of these dissecting procedures, all muscular attachments should be observed and, in all joints dissected, the articular surfaces and movements studied.

Summary of structures

1. Bones and bony landmarks
   - Carpals
   - Humerus
   - Lateral epicondyle of humerus
   - Medial epicondyle of humerus
   - Metacarpals
   - Olecranon
   - Radius
   - Scapula
   - Ulna
   - Ulna, coronoid process
   - Ulna, head
   - Ulna, styloid process

Table 21

Name	Origin	Insertion	Innervation	Action
Biceps brachii (see Table 16)				
Brachialis (see Table 16)				
Infraspinatus (see Table 19)				
Subscapularis (see Table 15)				
Supraspinatus (see Table 19)				
Teres major (see Table 19)				
Teres minor (see Table 19)				
Triceps brachii (see Table 19)				

2. Bursae
 Subscapular
 Tricipital
3. Capsule
 Articular
4. Disc
 Intraarticular, wrist
5. Joints
 Acromioclavicular
 Carpal
 Distal radioulnar
 Elbow
 Glenohumeral
 Interphalangeal
 Metacarpophalangeal
 Proximal radioulnar
 Wrist
6. Ligaments
 Annular
 Carpometacarpal
 Coracoacromial
 Coracoclavicular
 Coracohumeral
 Dorsal intercarpal
 Dorsal radiocarpal
 Inferior glenohumeral
 Intermetacarpal
 Middle glenohumeral
 Palmar intercarpal
 Palmar radiocarpal
 Palmar ulnocarpal
 Pisohamate
 Pisometacarpal
 Radial collateral, elbow
 Radial collateral, wrist
 Superior glenohumeral
 Ulnar collateral, elbow
 Ulnar collateral, wrist
7. Muscles
 Biceps brachii
 Biceps brachii, long head
 Brachialis
 Infraspinatus
 Subscapularis
 Supraspinatus
 Teres major
 Teres minor
 Triceps brachii
8. Surface
 Articular

Part Three
THORAX

DISSECTION AND REMOVAL OF THE ANTEROLATERAL HALF OF THE THORACIC CAGE: DISPOSITION OF THORACIC ORGANS IN SITU

Observation. Observe on the anterior and lateral aspects of the thorax (pp. 180 and 181) the jugular notch, sternum, sternal angle, xiphoid process, ribs 1 to 10, and costal cartilages. Study with particular attention the thoracic aperture (pp. 168 to 171) and the elements in it.

Palpation. Palpate the observed elements and then the intercostal spaces, costal arches, and the angle between the arches (intercostal angle, infrasternal angle).

Dissection. Remove completely the pectoralis major, pectoralis minor, serratus anterior, and subclavius muscles from the thorax.

From the thoracic attachment of the serratus anterior muscle down, dissect the external oblique muscle of the abdomen to the costal arch.

Identify by palpation the fifth or sixth intercostal space and dissect, in it, from the sternum posterolaterally, the

Table 22

Name	Origin	Insertion	Innervation	Action
Diaphragm	Xiphoid process, cartilages of lower six ribs, medial and lateral lumbocostal arches, and L1 to L3 vertebral bodies	Central tendon	Phrenic nerve	Active in inspiration
External intercostal	Lower border of rib, from tubercle to costal cartilage	Upper border of next inferior rib	Intercostal nerves	Active in respiration and speech
External oblique of abdomen	External surface of lower eight ribs	Iliac crest, anterior superior iliac spine, and (by way of aponeurosis) inguinal ligament and linea alba	Ventral rami of lower six thoracic spinal nerves	Stabilizes abdominal viscera; active in expiration; flexes and laterally flexes lumbar column; rotates trunk
Innermost intercostal	Internal surface of rib	Internal surface of next inferior rib	Intercostal nerves	Active in respiration and speech
Internal intercosal	Floor of costal groove, from sternal end to posterior costal angle	Upper border of next inferior rib	Intercostal nerves	Active in respiration and speech
Pectoralis major (see Table 15)				
Pectoralis minor (see Table 15)				
Serratus anterior (see Table 15)				
Subclavius (see Table 15)				
Transversus thoracis	Posterior surface of xiphoid process, sternum, and costal cartilage	Lower borders and internal surface of second to sixth ribs	Intercostal nerves	Depresses costal cartilages

external intercostal membrane and the external intercostal muscle. Remove the muscle.

Expose, in the same intercostal space, the superficial layer of the internal intercostal muscle. Remove this layer from the upper rib and deep to it find, from above downward, the posterior intercostal vein and artery and the intercostal nerve. Simultaneously expose the external aspect of the deep layer of the internal intercostal muscle, the innermost intercostal muscle. Repeat the dissection at several other intercostal levels.

Find the initial part of the internal thoracic artery, of the subclavian artery, and follow it through the superior thoracic aperture. Dissect, while in the aperture, the vagus and the phrenic nerves. Observe the pleural domes as they protrude through the aperture into the neck.

Cut ribs 1 to 9 on both sides of the thoracic cage, along the vertical line passing through the centers of the ribs, with a saw or bone shears. Before the first rib is cut, try to separate the parietal pleura from its deep aspect. Repeat the same procedure while cutting each subsequent rib.

After the section through the ribs is completed, separate the anterior half of the thoracic cage from the underlying thoracic organs. Cut the internal thoracic artery close to its origin. Reflect the anterior half of the cage anteriorly and observe the anterior attachment of the diaphragm to the internal aspects of several ribs. Cut the ninth, eighth, and seventh costal cartilages just above this attachment, proceeding from lateral to the anterior midpoint of the sternum. Remove the anterior half of the thoracic cage. In the course of this procedure the terminal branches of the internal thoracic artery—the musculophrenic and the superior epigastric—will be severed just above the level of the diaphragm.

Observe on the internal surface of the removed part of the thoracic cage the following structures: manubrium sterni, sternal body, costal cartilages, ribs, endothoracic fascia, innermost intercostal muscle, transversus thoracis muscle, internal thoracic blood vessels.

Summary of structures

1. Arteries
   Intercostal, posterior
   Internal thoracic
   Musculophrenic
   Subclavian
   Superior epigastric
2. Bones and bony landmarks
   Jugular notch
   Manubrium sterni
   Ribs
   Sternal angle
   Sternum
   Thoracic aperture
   Xiphoid process
3. Cartilage
   Costal
4. Membranes
   Endothoracic
   External intercostal
   Parietal pleura
   Pleural dome
5. Muscles
   Diaphragm
   External intercostal
   External oblique of abdomen
   Innermost intercostal
   Internal intercostal
   Pectoralis major
   Pectoralis minor
   Serratus anterior
   Subclavius
   Transversus thoracis
6. Nerves
   Intercostal
   Phrenic
   Vagus
7. Surface anatomy
   Costal arch
   Intercostal angle
   Intercostal space
   Thorax, anterior aspect
   Thorax, lateral aspect
8. Veins
   Intercostal, posterior
   Internal thoracic

VESSELS AND NERVES IN THE MEDIASTINUM

Observation. Observe (p. 236) the anterior costomediastinal recesses, the cardiac notch on the left lung, and the superior and inferior interpleural triangles. Observe, within the inferior triangle, the bare area of the pericardium that is not covered by the lung—the zone of absolute solidity on percussion. Observe, within the superior triangle, the connective tissue, remnants of the thymus, and the large blood vessels. Reflect somewhat the inferior part of the left anterior costomediastinal recess and observe that part of the pericardium which is covered only by a thin layer of the pulmonary tissue—the zone of relative solidity on percussion.

Restore the anterior half of the thoracic cage and study the superficial projections of the recesses, triangles, and zones.

Dissection. Reflect the anterior costomediastinal recesses laterally (p. 238) and dissect the anterior aspect of the pericardium, from the diaphragm up to the level of the superior vena cava, aorta, and the pulmonary trunk.

Find on each side of the neck the junction between the internal jugular and subclavian veins and dissect from there the left and right brachiocephalic veins (p. 240). Follow the veins to their junction, and from it expose the superior vena cava as it enters into the pericardium. Reflect the latter vein to the left, and behind it identify the azygos vein as it arches over the right pulmonary root to finally empty into the superior cava. Cut the azygos vein close to its termination.

Cut through the initial parts of the brachiocephalic veins while preserving, in the neck, the junction between the internal jugular and subclavian veins, and the terminal part of the thoracic duct on the left side.

Dissect the phrenic nerves from the neck to the diaphragm.

Follow the left vagus nerve from the neck as it crosses the anterior aspect of the arch of the aorta, and find the origin of the left recurrent laryngeal nerve, which passes inferior to and then ascends behind the aortic arch.

Follow the right vagus nerve across the anterior aspect of the right subclavian artery, and find the right recurrent laryngeal nerve arching inferior and then behind the artery. Proceed by following the right vagus nerve to the level where it descends posterior to the right bronchus.

Dissect the arch of the aorta from the ascending portion of the vessel to the left and posteriorly, over the left bronchus, to where it becomes continuous with the descending aorta. Isolate the three branches of the aorta, from right to left: the brachiocephalic trunk, which bifurcates into the right common carotid and the right subclavian arteries; the left common carotid artery; and the left subclavian artery.

Immediately to the left of the ascending aorta, and below the arch, expose the anterior aspect of the pulmonary trunk to its bifurcation. Dissect further the left pulmonary artery toward the left lung. Observe the ligamentum arteriosum, which extends from the left pulmonary artery to the aorta. Note the relationship of this ligament with the left recurrent laryngeal nerve. This ligament is an important arterial communication between the two vessels in fetal circulation.

Dissect the right pulmonary artery within the space limited by the ascending aorta and the right lung.

Dissect (p. 242) the pulmonary veins, starting from the pulmonary hili toward the pericardium, while reflecting the heart to the opposite side. Note the relationship of the heart to the neighboring organs. Project the anterior aspect of the heart on the anterior surface of the thoracic cage. Cut (p. 244) the pulmonary veins and arteries close to their exit from or their entry into the lung. Cut the aorta immediately to the left of the attachment of the ligamentum arteriosum and its three branches close to their origins from the arch. Starting from the apex of the heart, peel the pericardium from the diaphragm backward and to the right, as far as the inferior vena cava. Cut this vein horizontally between its opening in the diaphragm and its entry into the right atrium. Separate with the fingers the rest of the pericardium from the surrounding organs and remove it, together with the heart, from the thorax.

Summary of structures

1. Arteries
   - Aorta
   - Aortic arch
   - Ascending aorta
   - Brachiocephalic
   - Common carotid
   - Descending aorta
   - Pulmonary
   - Pulmonary trunk
   - Subclavian
2. Duct
   - Thoracic

Table 23

Name	Origin	Insertion	Innervation	Action
Diaphragm (see Table 22)				

3. Gland
   Thymus
4. Ligament
   Arteriosum
5. Muscle
   Diaphragm
6. Nerves
   Phrenic
   Recurrent laryngeal
   Vagus
7. Organs
   Apex of heart
   Bronchus
   Heart
   Lung
   Pulmonary hilum
   Pulmonary root
   Right atrium
8. Pericardium
   Bare area
   Zone of absolute solidity
   Zone of relative solidity
9. Pleural structures
   Anterior costomediastinal recess
   Cardiac notch
   Inferior interpleural triangle
   Superior interpleural triangle
10. Surface anatomy
    Projection of heart
    Projection of recesses
    Projection of triangles
    Projection of zones
11. Veins
    Azygos
    Azygos, arch
    Brachiocephalic
    Inferior vena cava
    Internal jugular
    Pulmonary
    Subclavian
    Superior vena cava

HEART

Dissection. Study the pericardium (pp. 246 and 248), which envelops the heart and the great vessels, and differentiate the transverse and oblique pericardial sinuses. Incise the pericardium from the apex of the heart, along the anterior aspect, toward and onto the ascending aorta. Reflect the flaps of the pericardium backward while observing the extent of the pericardial cavity. Around the great vessels study the line along which the parietal pericardium reflects to become continuous with the visceral layer. Remove completely the parietal pericardium along this line of reflection.

Study the surfaces, margins, base, and the apex of the heart. Cut the aorta horizontally about 1 cm above its origin from the left ventricle. Demonstrate the aortic valves (p. 258) and find the openings of the left and right coronary arteries in the respective aortic sinuses.

On the surface of the heart, outside the lumen of the aorta, identify the origins of both coronary arteries. Dissect the left coronary artery (p. 250) and all the branches on the heart. Dissect the right coronary artery (p. 252) in the same way. Expose the coronary sinus in the inferior part of the atrioventricular groove and dissect its main tributaries—the great cardiac vein, posterior vein of the left ventricle, middle cardiac vein, small cardiac veins, and the oblique vein of the left atrium.

Observe on the posterior aspect of the right atrium the sulcus terminalis, which extends vertically from the right side of the opening of the superior vena cava to the right side of the opening of the inferior vena cava. Open the right atrium, just to the left of the sulcus terminalis, from the opening of one to the opening of the other vena cava (p. 256). Clean the interior of the right atrium and study the structures in it—the right auricle, right atrioventricular orifice, openings of the inferior and superior venae cavae and the coronary sinus, crista terminalis, musculi pectinati, and the fossa ovalis. Open the coronary sinus longitudinally, from its orifice into the atrium to the left, into the atrioventricular, or the coronary, sulcus.

Cut the pulmonary veins close to their entry into the left atrium. Cut the posterior wall of the left atrium between the four pulmonary veins and remove it. Clean the interior of the left atrium and study the following structures in it: the fossa ovalis, the left atrioventricular orifice, the left auricle, and the openings of the pulmonary veins.

Cut the pulmonary trunk above and close to the origin from the right ventricle. Demonstrate the pulmonary semilunar valves (p. 259). Cut longitudinally the initial part of the pulmonary trunk between the two semilunar valves and proceed by cutting the wall of the right ventricle, along the right and inferior margin of the heart, as far as the apex (p. 254). Clean the interior of the right ventricle and study the following structures in it: the atrioventricular valves, chordae tendineae, papillary muscles, ridges and bridges, and the septomarginal trabecula (moderator band).

Incise longitudinally the initial part of the aorta between the two semilunar valves and proceed by cutting the musculature of the left ventricle along the left margin of the heart as far as the apex. Clean the interior of the left ventricle and study the following structures in it: the atrioventricular valves, chordae tendineae, papillary muscles, and the ridges and bridges.

Compare the thickness of the left and right ventricular walls. Remove the ventricular walls from the anterior aspect of the heart to demonstrate the interventricular septum. Remove the atrial walls from the posterior aspect of the heart to demonstrate the interatrial septum.

Study the conducting system of the heart.

Reconstruct the anterior thoracic wall and observe the projection of the heart and the cardiac chambers on it. Attempt to observe the same projections on a living person.

Summary of structures

1. Arteries
   Aorta
   Aortic sinuses
   Left coronary
   Pulmonary trunk
   Right coronary
2. Heart
   Apex
   Atrioventricular groove
   Base
   Conducting system
   Interatrial septum
   Interventricular septum
   Margins
   Opening of inferior vena cava
   Opening of superior vena cava
   Openings of pulmonary veins
   Sulcus terminalis
   Surfaces
   Ventricular walls
3. Left atrium
   Fossa ovalis
   Left atrioventricular orifice
   Left auricle
4. Left ventricle
   Bridge muscles
   Chordae tendineae
   Papillary muscles
   Ridge muscles
5. Pericardium
   Oblique sinus
   Parietal
   Pericardial cavity
   Reflection
   Transverse sinus
   Visceral
6. Right atrium
   Crista terminalis
   Fossa ovalis
   Musculi pectinati
   Opening of coronary sinus
   Right atrioventricular orifice
   Right auricle
7. Right ventricle
   Bridge muscles
   Chordae tendineae
   Papillary muscles
   Ridge muscles
   Septomarginal trabecula
8. Surface anatomy
   Projection of cardiac chambers
   Projection of heart
9. Valves
   Aortic semilunar
   Left atrioventricular
   Pulmonary semilunar
   Right atrioventricular
10. Veins
   Coronary sinus
   Great cardiac
   Middle cardiac
   Oblique of left atrium
   Posterior of left ventricle
   Pulmonary
   Small cardiac

TRACHEA, BRONCHI, AND LUNGS

Dissection. Dissect (p. 260) the anterior aspect of the trachea from the neck down to the bifurcation. Expose the main bronchi, from the bifurcation to the lungs. Separate the vagus nerves from the posterior aspect of the bronchi and trace them toward the esophagus. Complete the dissection of the arch of the aorta as it crosses over the left bronchus and of the arch of the azygos vein as it crosses over the right bronchus. While dissecting the trachea and bronchi, identify those lymph nodes which are closely related to the extrapulmonary airways—the bronchopulmonary, the tracheobronchial, the tracheal, and the anterior mediastinal. Separate the trachea and bronchi from the contents of the posterior mediastinum and study precisely their relationships.

Try to peel the parietal pleura from the thoracic cage backward toward the vertebral column while passing the hand between the costal surface of the lung and the thoracic cage. Likewise, peel the diaphragmatic part of the pleura from the diaphragm and note the pleural costodiaphragmatic recesses. Separate the mediastinal pleura from the contents of the posterior mediastinum backward toward the vertebral column.

Cut the trachea horizontally, about 5 cm superior to the bifurcation.

Remove both lungs, covered by the parietal pleura, together with the trachea and bronchi, from the thorax (pp. 262 and 263).

Incise the parietal pleura along the anterior costomediastinal and costodiaphragmatic recesses and observe the pleural cavity. Reflect the entire parietal pleura onto the mediastinal surface of the lung toward the line along which it reflects to become continuous with the visceral layer.

Observe on the costal surface of the right lung (p. 264) the oblique and horizontal fissures and the three lobes; on the mediastinal surface of the right lung (p. 266) observe the hilum and the impression of the arch of the azygos vein above and behind it. Study the relationships among the elements of the right pulmonary root at the hilum.

Observe on the costal surface of the left lung (p. 265) the oblique fissure, two lobes, the cardiac notch, and the lingula; on the mediastinal surface of the left lung (p. 267) note the hilum, the cardiac impression, and the impression of the descending aorta above and behind the hilum. Study the relationships among the elements of the left pulmonary root at the hilum

Cut both bronchi close to their pulmonary entry.

Study the tracheal walls and the carina in a horizontal section through the organ.

Dissect, following the main bronchus into the lung, the bronchi of the second and the third orders (pp. 268 and 270). This dissection should be done by removing the pulmonary tissue, arteries, and veins through the mediastinal surface of the lung. Keep all other surfaces of the lung intact. When the third-order (segmental) bronchi are exposed, study their relationship with the costal surface of the lung.

Reconstruct the anterior thoracic wall and observe the projections of the pulmonary and pleural structures on it. Attempt to observe the same projections on a living person.

Summary of structures

1. Airways
   Bronchus, first-order
   Bronchus, second-order
   Bronchus, third-order
   Carina
   Trachea
   Tracheal bifurcation
2. Artery
   Aortic arch
3. Bones
   Vertebral column
4. Lung
   Cardiac impression
   Cardiac notch
   Costal surface
   Hilum
   Horizontal fissure
   Impression of arch of azygos vein
   Impression of descending aorta
   Lingula
   Lobes
   Mediastinal surface
   Oblique fissure
   Pulmonary root
5. Lymph nodes
   Anterior mediastinal
   Bronchopulmonary
   Tracheal
   Tracheobronchial
6. Nerve
   Vagus
7. Organ
   Esophagus
8. Pleura
   Anterior costomediastinal recess
   Costodiaphragmatic recess
   Line of reflection
   Parietal
   Parietal, costal
   Parietal, diaphragmatic
   Parietal, mediastinal
   Pleural cavity
   Visceral
9. Surface anatomy
   Projection of lung
   Projection of pleural recesses
10. Vein
    Arch of azygos

POSTERIOR MEDIASTINUM

Dissection. Continue the dissection of the vagus nerves from where they cross the posterior aspects of the bronchi downward, as they approach the lateral aspect of the esophagus and divide into the anterior and posterior branches (p. 272). Follow the anterior branches from both sides, along the anterior aspect of the esophagus, until they join together to form the anterior vagal trunk.

Dissect the entire anterior and lateral surfaces of the esophagus from the neck down to the hiatus of the diaphragm. Observe the relationships of the esophagus to the other structures in the posterior mediastinum. Reflect the esophagus forward while simultaneously dissecting its posterior aspect. On the posterior aspect of this organ, complete the dissection of the posterior vagal branches until they join one another to form the posterior vagal trunk. Reflect the esophagus downward onto the diaphragm.

Continue the dissection of the arch and subsequently of the descending aorta downward, toward the aortic hiatus of the diaphragm. Observe the relationship of the aorta to the other structures in the posterior mediastinum. Expose from the aorta the initial parts of the posterior intercostal arteries on both sides. Remove gradually the posterior intercostal arteries from their origins and reflect the aorta downward onto the diaphragm.

Along the anterior aspect of the right half of the vertebral column dissect the azygos vein, from the diaphragm upward toward the arch of this vein (p. 274). To the left of the azygos vein and on the anterior aspect of the vertebral column complete the dissection of the thoracic duct. Trace the duct from the diaphragm upward, as far as its junction with one of the major veins at the root of the neck, on the left side. To the left of the thoracic duct, which has a midline position in the lower part of the mediastinum, and still on the anterior aspect of the vertebral column dissect the tributaries of the azygos vein—the hemiazygos and the hemiazygos accessory veins. Observe how the posterior intercostal veins from both sides contribute to the formation of the azygos venous system.

Expose, proceeding from the cervicothoracic (stellate) sympathetic ganglion downward, the thoracic sympathetic trunk, the ganglia, and the interganglionic communications (p. 276). From about the fifth sympathetic ganglion inferiorly, isolate the main sympathetic trunk and the roots of the splanchnic nerves.

Remove from within several intercostal spaces the internal intercostal membrane and muscle and expose the posterior intercostal vein and artery and the intercostal nerve. Follow the intercostal nerve as it courses posterior to the sympathetic trunk and expose the communications with the corresponding sympathetic ganglion—the gray and white rami communicantes.

Summary of structures

1. Arteries
   Aortic arch
   Descending aorta
   Intercostal, posterior
2. Bones
   Vertebral column
3. Duct
   Thoracic
4. Ganglia
   Cervicothoracic (stellate) sympathetic
   Thoracic sympathetic
5. Membrane
   Internal intercostal
6. Muscles
   Diaphragm
   Internal intercostal
7. Nerves
   Anterior vagal trunk
   Gray ramus communicans
   Intercostal
   Interganglionic communicating
   Posterior vagal trunk
   Splanchnic
   Sympathetic trunk
   Vagus
   Vagus, anterior branch
   Vagus, posterior branch
   White ramus communicans
8. Organs
   Bronchus
   Esophagus
9. Spaces
   Intercostal
   Posterior mediastinum
10. Veins
   Arch of azygos
   Azygos
   Hemiazygos
   Hemiazygos accessory
   Intercostal, posterior

Table 24

Name	Origin	Insertion	Innervation	Action
Diaphragm (see Table 22)				
Internal intercostal (see Table 22)				

Part Four
ABDOMEN

ANTERIOR ABDOMINAL WALL

Observation. Observe (pp. 180 and 181) the xiphoid process and the costal arches superiorly, the anterior superior iliac spine, inguinal sulcus, and the symphysis pubis inferiorly, and the limits of the anterior abdominal wall, the midline indentation of the linea alba, and the umbilicus.

Palpation. Palpate (p. 168) the upper and lower bony limits of the region, and then in a living subject the anterior margin of the liver, ascending and descending colons, and the spermatic cord where it passes through the superficial inguinal ring, just above the pubis.

Dissection (Fig. 9). The upper skin incision has already been made. Make the lower skin incision parallel to and 1 cm below the inferior limit of the region (the symphysis pubis—inguinal sulcus—anterior superior iliac spine) and proceed along the iliac crest posteriorly. Make a median incision from the tip of the xiphoid process down, around the umbilicus, and to the symphysis pubis. Reflect the skin from the median line laterally.

Dissect the superficial circumflex iliac and superficial epigastric arteries in the subcutaneous tissue, from the femoral artery toward the anterior superior iliac spine and the umbilicus, respectively. Reflect the arteries toward their origins.

Remove the fascia (p. 236) and expose the external oblique muscle of the abdomen and the attachments. Observe the two crura and, between them, the superficial inguinal ring with the spermatic cord in the male, or the round ligament in the female (p. 236). Demonstrate the transverse, intercrural fibers that unite both crura on the anterior aspect of the aponeurosis. Dissect the components of the spermatic cord apart and study their relationship.

Incise the external oblique muscle of the abdomen from the point where the two crura meet laterally and upward, following the direction of muscular fibers. Make a vertical incision through the muscle, from the same point up to the thoracic attachment of this muscle. Reflect the muscle, now divided into the three parts, down to the iliac crest and the inguinal ligament, up to the thorax, and medially. Note how the latter part, which is aponeurotic, fuses with the aponeurosis of the internal oblique muscle.

Expose the surface and the attachments of the internal oblique muscle of the abdomen (p. 238). Observe the inferior, free border from which a few muscular fibers, the cremaster muscle, arise to join the spermatic cord. Cut and subsequently reflect this muscle similarly to the way the external oblique muscle of the abdomen was done. Dissect, deep to the reflected internal oblique muscle of the abdomen, the deep circumflex iliac artery, a branch of the external iliac artery, and the iliohypogastric and ilioinguinal nerves, near the anterior third of the iliac crest.

Expose the surface and the attachments of the transverse abdominis muscle (p. 240) medially to the semilunar line, at which level the muscle becomes continuous with the aponeurosis.

Cut vertically through the middle of the anterior layer of the rectus abdominis sheath and reflect the parts laterally and medially to the corresponding borders of the muscle. Dissect the surface of the rectus abdominis muscle and observe several fibrous intersections. Identify the pyramidalis muscle as it crosses the anterior aspect of the lower part of the rectus abdominis muscle. Cut the rectus abdominis muscle horizontally through the middle and reflect the parts superiorly and inferiorly. In the process of the latter procedure observe the superior epigastric artery, a branch of the internal thoracic artery, on the deep aspect of the upper muscular part; and the inferior epigastric artery, a branch of the external iliac artery, on the deep aspect of the lower muscular part. Find, posterior to the rectus abdominis muscle, the arcuate line, which represents the free margin of the posterior layer of the rectus sheath. From the arcuate line downward and still posterior to the muscle, observe a part of the transversalis fascia.

Cut and subsequently reflect the transverse abdominis muscle in a way similar to how the oblique muscles were dissected away (p. 242). Expose on the entire surface of the anterior abdominal wall the transversalis fascia, which is fused with the parietal layer of the peritoneum. Cut horizontally the transversalis fascia, together with the parietal peritoneum, just below the umbilicus and vertically on each side from the anterior superior iliac spine to the corresponding costal arch.

Reflect the lower segment of the fascia inferiorly and observe the following, from medial to lateral on the deep

Fig. 9

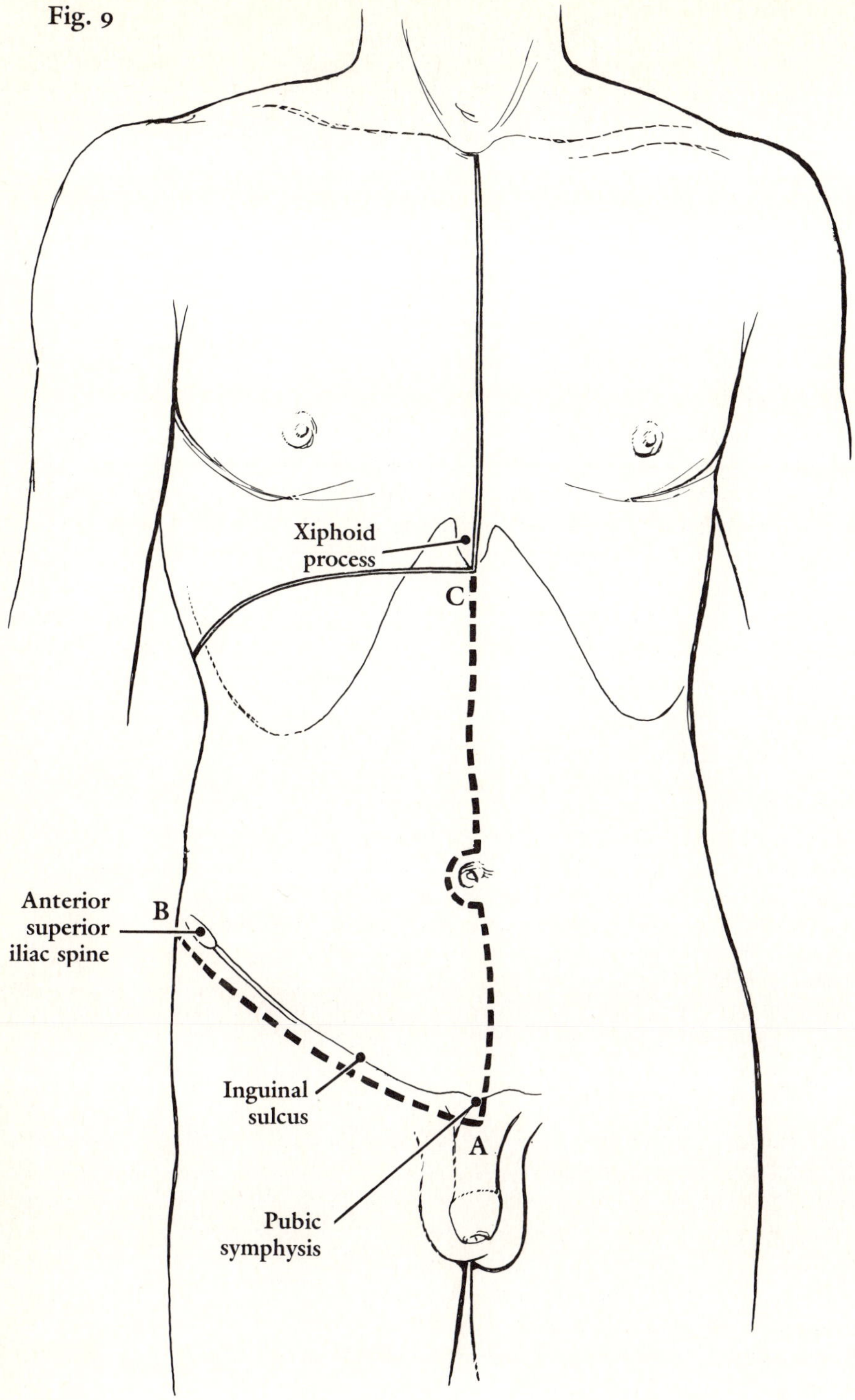

aspect of the reflected fascia: the median umbilical fold, the urachus; the supravesical fossa; the medial umbilical fold, the obliterated umbilical artery; the medial inguinal fossa; the lateral umbilical fold, the inferior epigastric vessels; the lateral inguinal fossa (p. 244).

Complete the dissection of the inferior epigastric artery to its origin.

In the course of this dissection identify gradually the intercostal (subcostal) and lumbar, iliohypogastric, and ilio-inguinal nerves as they travel from the posterior region, through the lateral abdominal wall, toward the anterior midline.

Reconstruct and subsequently study the organization of the inguinal canal and its content.

Table 25

Name	Origin	Insertion	Innervation	Action
Cremaster	Inguinal ligament; internal oblique and tranverse muscles of abdomen	Fascia of spermatic cord and tunica vaginalis	Genital branch of genito-femoral nerve	Elevates testis
External oblique of abdomen (see Table 22)				
Intrenal oblique of abdomen	Inguinal ligament, iliac crest, and thoracolumbar fascia	Inferior border of lower three ribs and (by way of aponeurosis) pecten pubis, linea alba, and cartilages of seventh, eighth, and ninth ribs	Ventral rami of lower six thoracic spinal nerves	Stabilizes abdominal viscera; active in expiration; flexes and laterally flexes lumbar column; rotates trunk
Pyramidalis	Symphysis pubis	Linea alba	Subcostal nerve	Tenses linea alba
Rectus abdominis	Symphysis pubis, pecten pubis, and linea alba	Fifth, sixth, and seventh costal cartilages	Ventral rami of lower six thoracic spinal nerves	Stabilizes abdominal viscera; active in expiration; flexes and laterally flexes lumbar column
Transverse of abdomen	Inguinal ligament, iliac crest, thoracolumbar fascia, and internal surface of lower costal cartilage	Pecten pubis and linea alba (by way of aponeurosis)	Ventral rami of lower six thoracic spinal nerves	Stabilizes abdominal viscera; active in expiration

Summary of structures

1. Aponeuroses
   Arcuate line
   Crura
   External oblique muscle of abdomen
   Internal oblique muscle of abdomen
   Linea alba
   Rectus sheath
   Rectus sheath, anterior layer
   Rectus sheath, posterior layer
   Semilunar line
   Superficial inguinal ring
   Transverse abdominis muscle
2. Arteries
   Deep circumflex iliac
   External iliac
   Femoral
   Inferior epigastric
   Internal thoracic
   Superficial circumflex iliac
   Superficial epigastric
   Superior epigastric
3. Bones and bony landmarks
   Anterior superior iliac spine
   Costal arch
   Iliac crest
   Pubis
   Symphysis pubis
   Xiphoid process
4. Canal
   Inguinal
5. Fasciae
   Deep
   Superficial
   Transversalis
6. Folds
   Lateral umbilical
   Medial umbilical
   Median umbilical
7. Fossae
   Lateral inguinal
   Medial inguinal
   Supravesical
8. Ligaments
   Inguinal
   Intercrural
   Round
9. Muscles
   Cremaster
   External oblique of abdomen
   Internal oblique of abdomen
   Pyramidalis
   Rectus abdominis
   Transverse of abdomen

10. Nerves
    Iliohypogastric
    Ilioinguinal
    Intercostal (subcostal)
    Lumbar
11. Peritoneum
    Parietal
12. Remnants
    Umbilical artery
    Urachus
13. Surface anatomy
    Anterior abdominal wall
    Anterior margin of liver
    Ascending colon
    Descending colon
    Inguinal sulcus
    Spermatic cord
    Umbilicus

DISPOSITION OF ORGANS IN THE ABDOMEN: LEFT GASTRIC AND HEPATIC ARTERIES

Cut the upper part of the transversalis fascia, together with the parietal peritoneum, vertically from the left side of the umbilicus up to the xiphoid process. Continue cutting the left segment of the divided fascia along the left costal arch and reflect it to the left.

Observation. Reflect the right segment of the fascia onto the right costal arch and observe the falciform ligament, which extends from the umbilicus to the anterior aspect of the liver. Observe simultaneously the round (teres) ligament, which extends from the umbilicus to the inferior aspect of the liver. Remove both ligaments from the umbilicus, together with the fascia and the parietal peritoneum, up to the right costal arch (p. 280).

Observe below the left costal arch the anterior aspect of the stomach with the greater and lesser curvatures (p. 282). Observe below the right costal arch the anterior margin of the liver, the fundus of the gallbladder, the falciform ligament, and the round (teres) ligament. Reflect this margin of the liver upward and expose the lesser omentum, which extends from the inferior surface of the liver to the lesser gastric curvature and to the duodenum. Identify the right, free border of the lesser omentum. Behind the content of this border, which consists of the bile duct, portal vein, and the hepatic artery, introduce the index finger through the epiploic foramen and into the lesser sac.

Follow the duodenum from the pyloric part of the stomach to the right until it becomes overlapped anteriorly by the colon.

Beginning from the greater curvature of the stomach and proceeding inferiorly, expose the entire surface of the greater omentum, and observe (1) the part of the greater omentum between this curvature and the transverse colon, the gastrocolic ligament, and (2) the inferior part of the greater omentum, which is freely suspended from the transverse colon. Reflect upward this free part of the greater omentum, together with the transverse colon, and observe first the entrance of the terminal part of the duodenum into the peritoneal cavity and then the duodenojejunal flexure, which is immediately to the left of the vertebral column. From the duodenojejunal flexure in the left abdominal compartment follow the horizontal flexures of the jejunum and, within the right abdominal compartment, the vertical flexures of the ileum. Observe the ileocolic junction at the end of the ileum. Expose the appendix behind the cecum. From the ileocolic junction follow the ascending colon up to the right flexure, the transverse colon from the right to the left flexture, the descending colon from the left flexure inferiorly, and subsequently the sigmoid colon.

Dissection. Expose the greater part of the lesser omentum by reflecting the stomach to the left and the anterior margin of the liver upward. Remove the anterior layer of the lesser omentum, starting from its right free border and proceeding toward the left while dissecting the bile ducts, portal vein, and hepatic artery. Remove the peritoneal covering from the inferior aspect of the gallbladder, and from the bladder dissect the cystic duct to its junction with the common hepatic duct. From this junction complete the dissection of the common hepatic duct superiorly, toward the point of union of the right and left hepatic ducts. Complete the dissection of the common bile duct inferiorly to the level of the duodenum.

Follow the hepatic artery, which was previously exposed in the right part of the lesser omentum, toward the liver, to its bifurcation into the right and left branches. Isolate the cystic artery, a branch of the right hepatic artery.

Expose subsequently the portal vein (which lies at a more posterior plane in relation to the biliary ducts and the hepatic artery), its bifurcation, and its right and left branches.

Cut the lesser omentum from the right to the left through the middle, and open the lesser sac. Follow the hepatic artery to the left as far as the celiac trunk by removing the peritoneum from the posterior wall of the lesser sac. Isolate the collateral branches of the hepatic artery—the right gastric and the gastroduodenal. In the course of this dissection note a great number of autonomic nerves from the celiac ganglion that accompany the hepatic artery, portal vein, and the bile duct.

Dissect simultaneously the left and right gastric arteries along the lesser gastric curvature from the celiac trunk and from the hepatic artery, respectively. Identify the anastomosis between the two gastric arteries.

Expose the beginning of the splenic artery from the celiac trunk.

Dissect the gastroduodenal artery from its origin downward, behind the duodenum, to the bifurcation into the right gastroepiploic and the superior pancreaticoduodenal arteries. By removing the anterior layer of the gastrocolic ligament along the greater curvature of the stomach, dissect the right gastroepiploic artery to the left until the anastomosis with the left gastroepiploic artery. Then follow the left gastroepiploic artery up to its origin from the splenic artery. The anastomosis between the gastroepiploic arteries, however, may not be noticeable.

Remove the gastrocolic ligament from the greater curvature and reflect it downward.

Study the walls and recesses of the lesser sac and the communication with the greater sac.

Summary of structures

1. Arteries
   Celiac trunk
   Cystic
   Gastroduodenal
   Hepatic
   Left gastric
   Left gastroepiploic
   Left hepatic
   Right gastric
   Right gastroepiploic
   Right hepatic
   Splenic
   Superior pancreaticoduodenal
2. Bones and bony landmarks
   Costal arch
   Vertebral column
   Xiphoid process
3. Ducts
   Bile
   Common bile
   Common hepatic
   Cystic
   Left hepatic
   Right hepatic
4. Fascia
   Transversalis
5. Ganglion
   Celiac
6. Ligaments
   Falciform
   Gastrocolic
   Round (teres)
7. Nerves
   Autonomic
8. Organs
   Appendix
   Ascending colon
   Cecum
   Colon
   Descending colon
   Duodenojejunal flexure
   Duodenum
   Gallbladder
   Gallbladder fundus
   Ileocolic junction
   Ileum
   Jejunum
   Left colic flexure
   Liver
   Right colic flexure
   Sigmoid colon
   Stomach
   Stomach, greater curvature
   Stomach, lesser curvature
   Stomach, pyloric part
   Transverse colon
9. Peritoneum
   Cavity
   Epiploic foramen
   Greater omentum
   Greater sac
   Lesser omentum
   Lesser sac
   Parietal
10. Surface anatomy
    Umbilicus
11. Veins
    Portal
    Portal, left branch
    Portal, right branch

SUPERIOR AND INFERIOR MESENTERIC ARTERIES

Dissection. Reflect the transverse colon superiorly and the flexures of the jejunum and ileum to the left. Locate a point 2 to 3 cm to the right of and below the duodenojejunal flexure (p. 295); from this point incise the parietal peritoneum obliquely to the ileocolic junction. Reflect the peritoneum to the right and left from the upper part of the incision and find the superior mesenteric artery (p. 284). Observe and subsequently remove a thick network of the autonomic nerves from around the artery. Dissect the artery further superiorly, in front of the third part of the duodenum and the uncinate process of the pancreas, and then behind the body of the pancreas all the way to the aorta. When the entire length of the superior mesenteric artery has been exposed, dissect the left branches—the jejunal and the ileal—from superior to inferior while removing the accompanying veins. In the course of this dissection observe the communications between each two adjacent branches and the formation of the arcades of the first order, the second order, and so on.

Expose the inferior pancreaticoduodenal artery from the first left branch of the superior mesenteric artery (occasionally it originates directly from the superior mesenteric artery) as far to the right as possible. Identify the anastomosis between this and the superior pancreaticoduodenal arteries.

Dissect the right branches of the superior mesenteric artey from superior to inferior: the middle colic, the right colic, and the ileocolic. Each of the right branches, before it reaches the intestine, divides into two branches that communicate with one another in the following manner: the left branch of the middle colic artery proceeds to the left colic flexure where it joins a branch of the inferior mesenteric artery; the right branch of the middle colic artery extends toward the right colic flexure, where it joins the ascending branch of the right colic artery; the descending branch of the right colic artery anastomoses with the ascending branch of the ileocolic artery; the descending branch of the ileocolic artery anastomoses with the last ileal artery. Expose the arterial branches to the cecum and the appendix. Dissect the straight arteries (p. 298) from the anastomoses of the aforementioned branches to the wall of the colon.

Reflect the jejunal and ileal coils to the right. Below and to the left of the duodenojejunal junction, incise the peritoneum vertically for a few centimeters, reflect the peritoneum, and expose the inferior mesenteric artery (p. 286). Dissect, to the left of the artery, the inferior mesenteric vein. Follow the artery up to the aorta and the vein approximately to the inferior border of the pancreas. Continue dissecting the artery and the vein inferiorly to the level of the pelvic inlet. Dissect from superior to inferor the collateral branches of the inferior mesenteric artery—the left colic and the sigmoid. Observe the bifurcations of these branches and the anastomoses of the descending and ascending counterparts along the descending and the sigmoid colons. The descending branch of the last sigmoid artery joins the superior rectal artery in the pelvis, the terminal branch of the inferior mesenteric artery.

Expose the straight arteries as you have done those along the superior mesenteric system.

Study the external morphology of the colon (p. 297): the appendices epiploicae, the teniae coli, and the haustrae. Tie the duodenojejunal flexure at two levels, close to one another, and cut the intestine between the ligatures. Tie in the same way and subsequently cut the terminal part of the sigmoid colon.

Remove gradually the branches of both mesenteric arteries and the mesentery from the jejunum, ileum, and colon, and then remove the intestine from the abdominal cavity.

Open a part of the jejunum (p. 293) and ileum (p. 294) longitudinally, clean the intestinal interior, and observe the configuration of the mucous membrane.

Open the cecum and the ascending colon (p. 296) by cutting longitudinally through their anterior aspects. Clean the interior. Observe the opening of the ileum in the large intestine, the ileocecal valve, the opening of the appendix, and the interior configuration of the colon.

Summary of structures

1. Arteries
   - Aorta
   - Arcades
   - Ileal branches
   - Ileocolic
   - Inferior mesenteric
   - Inferior pancreaticoduodenal
   - Jejunal branches
   - Left colic
   - Middle colic
   - Right colic
   - Sigmoid
   - Straight
   - Superior pancreaticoduodenal
   - Superior rectal
2. Bony landmark
   - Pelvic inlet
3. Nerves
   - Autonomic
4. Organs
   - Appendices epiploicae
   - Appendix
   - Ascending colon
   - Cecum

Colon
Descending colon
Duodenojejunal flexure
Duodenum
Haustrae
Ileocecal valve
Ileocolic junction
Ileum, flexures
Jejunum, flexures
Pancreas, body
Pancreas, uncinate process
Sigmoid colon
Teniae coli
Transverse colon

5. Peritoneum
   Mesentery
   Parietal
6. Veins
   Inferior mesenteric
   Superior mesenteric

DUODENUM, PANCREAS, LIVER, SPLEEN, SPLENIC ARTERY, AND CELIAC PLEXUS

Dissection. Dissect, starting from the pylorus, the anterior surface of the four parts of the duodenum to the duodenojejunal junction, and observe the relationship of each specific segment with the surrounding organs (p. 288). Within the concavity of the duodenum remove the parietal peritoneum from the anterior surface of the head of the pancreas, and proceed by removing it toward the left from the pancreatic body and the tail. Continue the dissection of the superior and inferior pancreaticoduodenal arteries. Observe the bifurcation of each of these arteries into an anterior and a posterior branch. Follow the two anterior branches, between the duodenum and the pancreas, and find the anastomosis between them. Observe their branches to the duodenum and the pancreas. The distribution of the posterior branches of these two arteries is similar on the posterior aspect of the two organs.

Cut the pancreas horizontally at the level of the pancreatic body midway between the upper and lower borders. Find the pancreatic duct by removing gradually the pancreatic tissue. Follow the duct and its tributaries to the left and to the right, as far as the extent of the gland (p. 290).

Continue the dissection of the common bile duct from behind the duodenum, as it extends posterior to the head of the pancreas, by removing the pancreatic tissue from the anterior surface of the gland. Follow this duct to its junction with the pancreatic duct.

Open vertically the anterior wall of the second duodenal part and find in its interior the opening of the pancreaticobiliary duct, the duodenal papilla.

Dissect the splenic artery from the celiac trunk to the left and observe its branches—the pancreatic and the left gastroepiploic. Complete the dissection of the splenic artery to the entry into the spleen—the splenic hilum.

Cut vertically through the pylorus to separate the duodenum from the stomach. Cut the common bile duct above the duodenum, the superior and inferior pancreaticoduodenal arteries close to their origins, and all pancreatic branches close to their origins from the splenic artery; subsequently remove the duodenum with the pancreas from the abdominal cavity.

Dissect the splenic vein from the spleen to the right. Complete the dissection of the superior and inferior mesenteric veins superiorly, and find the beginning of the portal vein, which results from their junction with the splenic vein.

Study the position and relationships of the liver. Separate with hands the anterior aspect of the liver from the inferior surface of the diaphragm. Cut the falciform ligament horizontally and trace its separated layers, the superior coronary ligaments, along the superior aspect of the liver. Reflect the liver downward. Identify by palpation the hepatic veins on the posterior aspect of the liver and separate them, together with the liver, from the inferior vena cava. Remove the liver from the abdominal cavity, after the inferior coronary ligaments have been separated from the diaphragm, and study its surfaces, margins, lobes, and the peritoneum (pp. 299 to 301). Open the gallbladder and observe the interior configuration.

Reflect the stomach upward and expose the right surface of the spleen. Observe the position and the relationships of the spleen. Cut the splenic artery and vein close to the splenic hilum and remove the spleen from the abdominal cavity. Study the surfaces, margins, and the impressions of the spleen (p. 302) and the relationships of this organ with the thoracic skeleton (p. 303). Cut the spleen horizontally and study the interior, the white and red pulps.

Expose the anterior aspect of the celiac ganglion around the celiac trunk. Observe in the thorax, behind the esophagus, the posterior vagal trunk. Follow this trunk and the

Table 26

Name	Origin	Insertion	Innervation	Action
Diaphragm (see Table 22)				

splanchnic nerves through the diaphragm and farther down to the celiac ganglion.

Cut the esophagus horizontally, just below the diaphragm. Cut through the origins of the left and right gastric and gastroepiploic arteries. Remove the stomach from the abdominal cavity. Open the stomach by cutting the wall along the greater curvature. Study the mucous membrane and observe the gastric muscular layers (p. 292).

Summary of structures

1. Arteries
   - Celiac trunk
   - Inferior pancreaticoduodenal
   - Left gastric
   - Left gastroepiploic
   - Right gastric
   - Right gastroepiploic
   - Splenic
   - Superior pancreaticoduodenal
2. Bones
   - Thoracic skeleton
3. Ducts
   - Common bile
   - Pancreatic
   - Pancreaticobiliary
4. Ganglion
   - Celiac
5. Ligaments
   - Falciform
   - Inferior coronary
   - Superior coronary
6. Muscle
   - Diaphragm
7. Nerves
   - Posterior vagal trunk
   - Splanchnic
8. Organs
   - Duodenal papilla
   - Duodenojejunal junction
   - Duodenum
   - Esophagus
   - Gallbladder
   - Liver
   - Liver, lobes
   - Liver, margins
   - Liver, surfaces
   - Pancreas
   - Pancreas, body
   - Pancreas, head
   - Pancreas, tail
   - Pancreas, tissue
   - Pylorus
   - Spleen
   - Spleen, hilum
   - Spleen, impressions
   - Spleen, margins
   - Spleen, red pulp
   - Spleen, surfaces
   - Spleen, white pulp
   - Stomach
9. Peritoneum
   - Parietal
10. Veins
    - Hepatic
    - Inferior mesenteric
    - Inferior vena cava
    - Portal
    - Splenic
    - Superior mesenteric

SUPRARENAL GLAND, KIDNEY, URETER, AORTA AND BRANCHES, INFERIOR VENA CAVA AND TRIBUTARIES

Dissection. Expose the inferior vena cava as it passes through the diaphragm, and the aorta between the two crura of the diaphragm. Observe the openings of the hepatic veins on the anterior wall of the inferior vena cava (p. 304).

Dissect the inferior phrenic arteries from their origins, which are in a position superior to that of the celiac trunk, as far as their entry into the diaphragm. Observe the collateral branches of these arteries (the superior suprarenal arteries), which extend laterally and inferiorly to the suprarenal glands. Dissect the inferior phrenic veins from the diaphragm to the inferior vena cava.

Complete the dissection of the celiac trunk and its branches.

Dissect the middle suprarenal arteries from the aorta, at

the level of or just below the origin of the celiac trunk, to the suprarenal glands. Remove the thin fascia from the anterior surface of the suprarenal glands, and observe their shapes, positions, and relationships. Dissect the additional suprarenal veins from the glands to the inferior vena cava. Expose the renal veins from the inferior vena cava to the kidneys. By reflecting subsequently the renal veins upward or downward, expose the renal arteries, which are in a position posterior to the veins, from the aorta to the kidneys. Observe how the arteries divide into many branches as they approach the renal hilum, and identify their inferior suprarenal branches.

Remove the thin fascia, together with the adipose capsule, from the anterior surfaces of the kidneys. Dissect the ureter from the renal hilum, where it is posterior to the arteries and veins, downward for about 2 cm until it is joined by the gonadal (testicular or ovarian) artery and vein. From this point follow the gonadal arteries up to their origins and the gonadal veins up to the inferior vena cava on the right side and the renal vein on the left side. Follow the gonadal vessels together with the ureter, down into the pelvic inlet.

Complete the dissection of the aorta and the inferior vena cava down to their bifurcations. Continue the dissection along the common iliac vessels, the bifurcations of the iliac vessels, and the external iliac vessels down to the level of the inguinal ligament.

Expose the median sacral artery from the posterior aspect of the aortic bifurcation along the pelvic surface of the sacrum.

Remove the suprarenal glands by cutting all arteries and veins close to the glandular surfaces.

Complete the dissection of the posterior surface of the kidney (p. 306). Cut the ureters about 2 cm peripheral to the renal hilum and the renal arteries and veins about 1 cm peripheral to the same hilum. Remove the kidneys. Cut one kidney frontally through the middle (p. 310). Wash the exposed surfaces and study the interior: hilum, renal sinus, ureter, renal pelvis, major calices, minor calices, medulla, pyramids, columns, cortex, capsule. Proceed, on the second kidney (p. 311), from the posterior aspect by removing the renal tissue and vessels and by exposing the posterior aspect of the renal pelvis and the major and minor calyces. Next cut this kidney horizontally at several levels and study the sections.

Cut the inferior vena cava and the aorta below their diaphragmatic openings. Remove all of their dissected branches. Reflect the inferior vena cava and the aorta gradually forward and observe the lumbar arteries and veins arising from the posterior aspect of the aorta or joining the posterior aspect of the inferior vena cava. Cut the common iliac vessels close to their bifurcations and remove them, together with the aorta and the inferior vena cava, from the posterior abdominal wall.

Expose the sympathetic trunks, the ganglia, and the interganglionic connections, along the lumbar vertebral column, down to the pelvic inlet.

Summary of structures

1. Arteries
   - Aorta
   - Celiac trunk
   - Common iliac
   - External iliac
   - Gonadal
   - Inferior phrenic
   - Inferior suprarenal
   - Lumbar
   - Median sacral
   - Middle suprarenal
   - Ovarian
   - Renal
   - Superior suprarenal
   - Testicular
2. Bones and bony landmarks
   - Pelvic inlet
   - Sacrum
   - Vertebral column
3. Fasciae
   - Renal, adipose
   - Suprarenal
4. Ligament
   - Inguinal
5. Muscles
   - Crura of diaphragm
   - Diaphragm
6. Nerves
   - Sympathetic ganglion
   - Sympathetic interganglionic connections
   - Sympathetic trunk

Table 27

Name	Origin	Insertion	Innervation	Action
Diaphragm (see Table 22)				

7. Organs
   Capsule of kidney
   Column of kidney
   Kidney
   Major calix
   Medulla of kidney
   Minor calix
   Pyramid of kidney
   Renal hilum
   Renal pelvis
   Renal sinus
   Suprarenal gland
   Ureter

8. Veins
   Common iliac
   External iliac
   Gonadal
   Hepatic
   Inferior phrenic
   Inferior vena cava
   Lumbar
   Ovarian
   Renal
   Suprarenal
   Testicular

DIAPHRAGM, POSTERIOR ABDOMINAL WALL, AND LUMBAR PLEXUS

Dissection. Start the dissection of the inferior aspect of the diaphragm (p. 318) by cleaning the central tendon. Expose the sternal part of the diaphragm from the central tendon to the sternum and the xiphoid process. Dissect the attachment of the diaphragm laterally and posteriorly all around the costal cartilages and ribs 7 to 11, the costal part of the diaphragm. Observe the space between the sternal and costal parts of the muscle and the superior epigastric artery passing through that space. Dissect the lumbar part of the diaphragm in three stages (p. 312):

1. Locate by palpation the tip of the twelfth rib and the L1 transverse process. Dissect between these two points the lateral arcuate ligament and that part of the diaphragm which is attached to it. Observe the potential space between the lumbar and costal parts of the diaphragm.

2. Dissect from the tip of the L1 transverse process to the lateral aspect of the L1 vertebral body the medial arcuate ligament and that part of the diaphragm which is attached to it.

3. Dissect on the right half of the anterior surface of the L1 to L3 vertebral bodies, and on the left side of the L1 and L2 vertebral bodies, the lumbar attachments of the diaphragmatic crura. Follow the crura upward and observe how they intersect to form the aortic hiatus of the diaphragm—the median arcuate ligament. Observe to the left and in front of the aortic opening how the muscular parts of the crura encircle the esophageal hiatus of the diaphragm. Complete the dissection of the crura as far as the central tendon.

Dissect the psoas muscle (p. 314) lateral to the lumbar vertebral bodies, starting from the level of the medial arcuate ligament. Proceed downward and expose the entire anterior aspect of the muscle to the level of the inguinal ligament. Identify the genitofemoral nerve on the anterior aspect of the muscle. Follow the nerve down to the bifurcation into the genital branch, which joins the spermatic cord, and the femoral branch. Expose also the entire lateral aspect of the psoas muscle down to the inguinal ligament.

Find the subcostal nerve, the ventral ramus of the T12 spinal nerve, as it passes beneath the lateral arcuate ligament, and dissect the nerve lateroinferiorly.

Find the ventral ramus of the L1 spinal nerve as it emerges from beneath the posterior border of the psoas muscle at the level between the L1 and L2 vertebral bodies. Follow the nerve lateroinferiorly, parallel to the subcostal nerve, to the bifurcation into the iliohypogastric and ilioinguinal nerves.

Find the lateral femoral cutaneous nerve as it emerges from beneath the posterior border of the psoas muscle, and follow the nerve to a point 2 cm medial and inferior to the anterior superior iliac spine.

Expose along the lateral aspect of the psoas muscle, below the level of the emergence of the lateral femoral cutaneous nerve, the femoral nerve as far down as the muscle is dissected.

Reflect the lower part of the psoas muscle laterally and find the obturator nerve along the medial border of the muscle. Subsequently identify, medial and slightly posterior to the obturator nerve, the lumbosacral trunk.

Dissect the anterior surface of the quadratus lumborum muscle, lateral to the psoas muscle, from the lateral arcuate ligament down to the iliac crest.

Expose, lateral to the quadratus lumborum muscle, the deep aspect of the lumbosacral aponeurosis and the transverse abdominis muscle.

Dissect the iliacus muscle and observe how it fuses with the psoas muscle in the iliac fossa.

Cut horizontally through the psoas muscle just above its junction with the iliacus muscle (p. 316). Dissect the branches of the lumbar plexus (the iliohypogastric, ilioinguinal, genitofemoral, lateral femoral cutaneous, obturator, and femoral) toward the spinal cord by removing piecemeal the upper part of the psoas muscle. Expose the components of the lumbar plexus—the ventral (anterior) rami of the lumbar nerves 1, 2, 3, and 4.

Table 28

Name	Origin	Insertion	Innervation	Action
Diaphragm (see Table 22)				
Iliacus	Superior two thirds of iliac fossa, iliac crest, sacroiliac and iliolumbar ligaments, and sacrum	Lesser trochanter of femur	Femoral nerve	Flexes thigh
Psoas major	Anterior surface of lumbar transverse processes and bodies of T12 and lumbar vertebrae	Lesser trochanter of femur	Ventral rami of L1 to L3 spinal nerves	Flexes thigh
Psoas minor	T12 and L1 vertebral bodies	Pecten pubis, iliopectineal eminence, and iliac fascia	Ventral ramus of L1 spinal nerve	Flexes trunk
Quadratus lumborum	Iliolumbar ligament, iliac crest, and transverse processes of upper four lumbar vertebrae	Last rib	Ventral rami of T12 to L3 spinal nerves	Active in inspiration; laterally flexes and extends lumbar column
Transverse of abdomen (see Table 25)				

Summary of structures

1. Aponeurosis
   Lumbosacral
2. Artery
   Superior epigastric
3. Bones and bony landmarks
   Anterior superior iliac spine
   Costal cartilage
   Iliac crest
   Iliac fossa
   Rib
   Sternum
   Transverse process of L1
   Vertebral body
   Xiphoid process
4. Central nervous system
   Spinal cord
5. Cord
   Spermatic
6. Ligaments
   Inguinal
   Lateral arcuate
   Medial arcuate
   Median arcuate
7. Muscles
   Diaphragm
   Diaphragm, aortic hiatus
   Diaphragm, central tendon
   Diaphragm, costal part
   Diaphragm, crura
   Diaphragm, esophageal hiatus
   Diaphragm, lumbar part
   Diaphragm, sternal part
   Iliacus
   Psoas
   Quadratus lumborum
   Transverse of abdomen
8. Nerves
   Femoral
   Genitofemoral
   Genitofemoral, femoral branch
   Genitofemoral, genital branch
   L1
   L1, iliohypogastric
   L1, ilioinguinal
   L1, ventral ramus
   L2, ventral ramus
   L3, ventral ramus
   L4, ventral ramus
   Lateral femoral cutaneous
   Lumbosacral trunk
   Obturator
   Subcostal
9. Plexus
   Lumbar

Part Five
PELVIS

INTERNAL ILIAC ARTERY AND SUPERIOR ASPECT OF PELVIC ORGANS

MALE

Observation. Observe (p. 322) from above the relief of the urinary bladder and behind it the rectum, both of which are covered by the parietal peritoneum in the pelvic inlet. Observe between these two organs the peritoneal (recto-vesicle) pouch. Locate in the pelvis the inferior mesenteric artery, the internal iliac artery, the urachus, the umbilical artery, the ductus deferens, the ureters, and the gonadal vessels.

Dissection. Remove (p. 326) the parietal peritoneum from the superior aspect of the urinary bladder backward. Follow the ductus deferens from the deep inguinal ring backward, and simultaneously follow the ureters to the bladder. Note how the ductus deferens crosses the terminal part of the ureter medially, and find where the ureters enter the wall of the urinary bladder.

Continue removing the peritoneum from the posterior aspect of the bladder while following the ductus deferens as far downward as possible.

Remove the peritoneum completely from the anterior aspect of the rectum. Reflect the rectum anteriorly and continue dissecting the inferior mesenteric artery over the posterior intestinal aspect. Expose the bifurcation of the artery into the superior rectal branches and follow the branches as far downward as possible.

Isolate the umbilical artery from the internal iliac artery and follow it to the lateral aspect of the urinary bladder. Display the branches of this artery, the artery of the ductus deferens, and the superior vesical artery. The latter artery ascends to the superior aspect of the urinary bladder.

Follow the obturator nerve along the lateral wall of the pelvis and note a branch of the internal iliac artery, the obturator artery, approaching it.

Display the origins of the superior and inferior gluteal and the internal pudendal arteries. Reflect the urinary bladder medially, and from the inferior gluteal artery (or sometimes from the internal pudendal artery) to the inferior aspect of the urinary bladder expose the inferior vesical artery and its branches to the prostate. Display the middle rectal artery from the internal iliac artery to the middle portion of the rectum. Palpate the prostate gland beneath the urinary bladder.

Reflect the anterior part of the urinary bladder backward and demonstrate the prevesical space. Dissect the pubovesical ligament in this space. Cut through the wall of the bladder from the apex in the median plane; observe the openings of the ureters and the urethra in the bladder and the trigone, an area delimited by these three openings.

FEMALE

Observation. Observe (p. 324) from above the relief of the urinary bladder in front and the uterus and rectum behind it, all of which are covered by the parietal peritoneum in the pelvic inlet. Observe, between the urinary bladder and the uterus, the vesicouterine peritoneal pouch, and between the uterus and the rectum the rectouterine peritoneal pouch. Note the double peritoneal layer, the broad ligament, extending from the uterus to the lateral wall of the pelvis. Find the round ligament, which extends inside the broad ligament from the uterus forward and laterally into the deep inguinal ring. Locate in the pelvis the inferior mesenteric artery, internal iliac artery, gonadal artery, umbilical artery, ureters, and the urachus.

Dissection. Remove (p. 328) the parietal peritoneum successively from the superior aspect of the urinary bladder, the superior aspect of the uterus, and toward the anterior aspect of the rectum. Remove the two peritoneal sheaths of the broad ligament and expose the uterine tube, which extends from the uterus laterally toward the ovary. Demonstrate the fimbriae of the uterine tube. Complete the dissection of the ovarian artery to the ovary, and simultaneously dissect the ureters to the urinary bladder.

Reflect the rectum anteriorly and dissect the inferior mesenteric artery over its posterior aspect as far as the bifurcation into the superior rectal branches. Follow the latter vessels downward.

Dissect the internal iliac artery from its origin and isolate its branch, the umbilical artery, which proceeds to the lateral aspect of the urinary bladder. Demonstrate the superior vesical branches of the latter artery to the superior aspect of the urinary bladder.

Dissect the uterine artery from the internal iliac artery as it passes first along the inferior margin of the broad ligament to the lateral aspect of the uterus. It then ascends along the lateral uterine wall toward the initial part of the uterine tube. The uterine artery finally accompanies the uterine tube all the way to the ovary. Isolate several small branches (the vaginal arteries) from that part of the uterine artery which is contained in the broad ligament.

Dissect the middle rectal branch of the internal iliac artery to the middle portion of the rectum.

Follow the obturator nerve along the lateral wall of the pelvis and note a branch of the internal iliac artery, the obturator artery, approaching it.

Expose the origins of the superior and inferior gluteal and the internal pudendal arteries from the internal iliac artery.

Reflect the uterus anteriorly and the rectum posteriorly, and beneath the uterus observe the posterior wall of the vagina. Subsequently reflect the uterus posteriorly and demonstrate the anterior aspect of the uterus and the vagina.

Reflect the apex of the urinary bladder posteriorly and demonstrate the prevesical space, which extends anteriorly to the pubic symphysis. Dissect the pubovesical ligament in this space.

Cut through the wall of the urinary bladder in the median plane; observe the openings of the ureters and the urethra in the bladder and the trigone, an area delimited by these three openings.

NOTE. In the course of this dissection all students are advised to follow, in addition to their own dissection, the dissection of a cadaver of the opposite sex.

If the pelvic organs are removed from just above the level of the pelvic diaphragm, the attachments, position, parts, and structure of the levator ani muscle can be appreciated (pp. 330 and 332).

Summary of structures

1. Arteries
   Ductus deferens
   Gonadal
   Inferior gluteal
   Inferior mesenteric
   Inferior vesical
   Inernal iliac
   Internal pudendal
   Middle rectal
   Obturator
   Ovarian
   Superior gluteal
   Superior rectal
   Superior vesical
   Umbilical
   Uterine
   Vaginal
2. Bony landmark
   Pelvic inlet
3. Diaphragm
   Pelvic
4. Ligaments
   Pubovesical
   Round
5. Muscle
   Levator ani
6. Nerve
   Obturator
7. Organs
   Apex of urinary bladder
   Ductus deferens
   Fimbriae
   Ovary
   Prostate
   Rectum
   Trigone of urinary bladder
   Urachus
   Ureter
   Ureteral opening
   Urethral opening
   Urinary bladder
   Uterine tube
   Uterus
   Vagina
8. Peritoneum
   Broad ligament
   Deep inguinal ring
   Parietal
   Prevesical space
   Rectouterine pouch
   Rectovesical pouch
   Vesicouterine pouch

Table 29

Name	Origin	Insertion	Innervation	Action
Levator ani (with coccygeus)	Pelvic surface of pubis, obturator fascia, spine of ischium, and sacrospinous ligament	Perineal body, median raphe, pelvic viscera, margin of coccyx, and side of S5 vertebra	Branch of S4 nerve	Supports pelvic viscera; constricts rectum and vagina

MUSCLES OF THE BACK AND GLUTEAL REGION

Observation. Observe (p. 335) the median sulcus in the lumbar region along the vertebral spines; it is limited by the two prominent muscular columns, which correspond to the erector spinae muscles. Further observe the iliac crest and below it the relief of the gluteal musculature, which extends as far down as the gluteal sulcus.

Palpation. Palpate (p. 169) the spines of the several lower thoracic and the lumbar vertebrae, sacrum, coccyx, eleventh and twelfth ribs, iliac crest, sacroiliac joints, posterior superior and posterior inferior iliac spines, ischial spine, sciatic notches, ischial tuberosity, erector spinae muscles, and the gluteal musculature.

Dissection of the muscles of the back. Incise the skin along the iliac crest from lateral to medial. Reflect it to the

Fig. 10

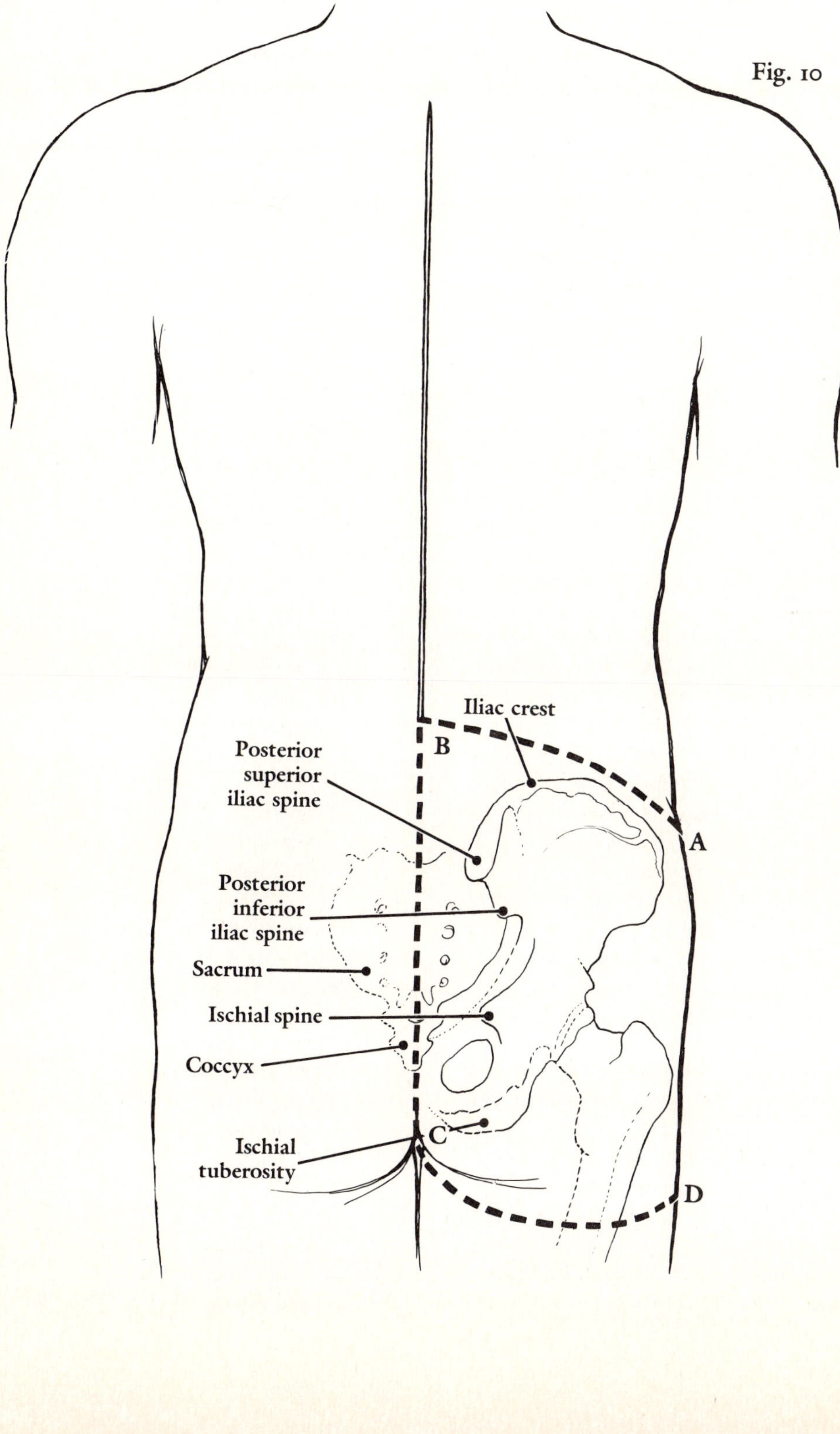

median plane and subsequently remove it. Expose the entire surface of the latissimus dorsi muscle as far as the vertebral column and the iliac crest.

Complete the dissection of the posterior segment of the external oblique muscle of the abdomen backward, to where the latissimus dorsi muscle covers it. Observe the lumbar triangle, which is limited by the latissimus dorsi muscle medially, the external oblique muscle of the abdomen laterally, and the iliac crest inferiorly. Remove the latissimus dorsi muscle and the external oblique muscle of the abdomen from their origins.

Expose the posterior inferior serratus muscle from the vertebral column to the last ribs (p. 338).

Complete the dissection of the posterior segment of the internal oblique muscle of the abdomen.

About 4 cm lateral and parallel to the median plane cut the aponeurosis of the erector spinae muscle and of the posterior inferior serratus muscle from the iliac crest up into the neck. Remove the aponeurosis and the posterior inferior serratus muscle and expose the posterior aspect of the erector spinae muscle. Observe the three divisions of the erector spinae muscle: the spinalis medially, the longissimus in the middle, and the iliocostalis laterally.

Dissection of the gluteal region. Incise the skin vertically (Fig. 10) along the median plane from the iliac crest down toward the anus. Proceed from the anus laterally and downward about 2 to 3 cm below and parallel to the gluteal sulcus. Reflect the skin laterally.

Cut the subcutaneous tissue from the sacrum obliquely and lateroinferiorly through the middle of the region, as deep as the surface of the gluteus maximus muscle. Remove the subcutaneous tissue and expose the entire surface of the muscle (p. 336). Using the hands, peel the deep surface of the muscle from the next muscular layer and then cut the muscle, perpendicular to the direction of the muscular fibers, through the middle. Reflect the medial part of the muscle medially while removing it carefully from the sacrotuberous ligament. Reflect the other part of the muscle laterally and observe a bursa on the deep muscular surface. In the course of this dissection note the inferior gluteal nerve and blood vessels entering the muscle through the deep surface. Cut the nerve and the vessels close to their entry into the muscle. Expose the sacrotuberous ligament (p. 338).

Locate precisely by palpation the posterior inferior iliac spine and the ischial spine; from in between these two points dissect the piriformis muscle to the greater trochanter. Note the superior gluteal nerve and blood vessels above the medial part of this muscle and dissect them into the pelvis as far as possible. Dissect from lateral to medial, along the inferior border of the piriformis muscle, the sciatic and posterior femoral cutaneous nerves down to the level of the skin incision, the inferior gluteal nerve and vessels, and the internal pudendal blood vessels together with the pudendal nerve.

Expose the sacrospinous ligament from the sacrum to the ischial spine, in front of or deep to the pudendal neurovascular bundle.

Dissect, in the upper part of the gluteal region, above the piriformis muscle, the gluteus medius muscle from the hip bone to the greater trochanter. Remove this muscle (p. 340).

Expose the gluteus minimus muscle while following the superior gluteal neurovascular bundle.

In the lower part of the gluteal region, below and parallel to the piriformis muscle, dissect each of the following four muscles to the greater trochanter: the gemellus superior from the ischial spine, the obturator internus from the lesser sciatic notch, the gemellus inferior from the ischial tuberosity, and the quadratus femoris from the ischial tuberosity.

The obturator internus muscle, which arises from the deep aspect of the obturator membrane in the pelvis, curves around the lesser sciatic notch and then proceeds through the gluteal region to the greater trochanter. To demonstrate its bursa, cut the gluteal part of this muscle vertically through the middle and reflect the medial part away from the lesser sciatic notch. Observe the bursa on the deep muscular aspect.

Summary of structures

1. Aponeurosis
   Erector spinae
2. Arteries
   Inferior gluteal
   Internal pudendal
   Superior gluteal
3. Bones and bony landmarks
   Coccyx
   Greater sciatic notch
   Greater trochanter
   Hip
   Iliac crest
   Ischial spine
   Ischial tuberosity
   Lesser sciatic notch
   Posterior inferior iliac spine
   Posterior superior iliac spine
   Ribs
   Sacrum
   Vertebral column
   Vertebral spines
4. Bursae
   Gluteus maximus
   Obturator internus
5. Joint
   Sacroiliac
6. Ligaments
   Sacrospinous
   Sacrotuberous

Table 30

Name	Origin	Insertion	Innervation	Action
Erector spinae (iliocostalis, longissimus, spinalis)	Median sacral crest, spines T11 to L5 vertebrae, supraspinous ligaments, iliac crest, and lateral sacral crest	Angle of ribs, transverse processes of C1 to T12 vertebrae, mastoid process, ligamentum nuchae, and spines of C7 to L2 vertebrae	Dorsal rami of cervical, thoracic, and lumbar spinal nerves	Extends head; rotates face to side of action; extends and laterally flexes vertebral column
External oblique of abdomen (see Table 22)				
Gemellus inferior	Tuberosity of ischium	Greater trochanter of femur	Nerve to quadratus femoris	Abducts and laterally rotates thigh
Gemellus superior	Spine of ischium	Greater trochanter of femur	Nerve to obturator internus	Abducts and laterally rotates thigh
Gluteus maximus	Posterior gluteal line of ilium, aponeurosis of erector spinae muscle, sacrum, coccyx, sacrotuberous ligament, and fascia of this muscle	Iliotibial tract, fascia lata, and gluteal tuberosity of femur	Inferior gluteal nerve	Abducts, extends, and laterally rotates thigh; active in walking and climbing
Gluteus medius	Between iliac crest and posterior gluteal line and anterior gluteal line	Greater trochanter of femur	Superior gluteal nerve	Abducts and medially rotates thigh
Gluteus minimus	Between anterior and inferior gluteal lines	Greater trochanter of femur	Superior gluteal nerve	Abducts and medially rotates thigh
Internal oblique of abdomen (see Table 25)				
Latissimus dorsi (see Table 15)				
Obturator internus	Inferior ramus of pubis, ramus of ischium, pelvic surface of hip bone from greater sciatic foramen to obturator foramen, and obturator membrane	Greater trochanter of femur	Nerve to obturator internus	Extends and abducts thigh
Piriformis	Anterior surface of sacrum and gluteal surface of ilium	Greater trochanter of femur	Branches from L5 to S2 spinal nerves	Abducts and laterally rotates thigh
Posterior inferior serratus	Spines of T11 to L2 vertebrae	Lower four ribs	Ventral rami of T9 to T12 spinal nerves	Pulls ribs posteroinferiorly
Quadratus femoris	Ischial tuberosity	Greater trochanter of femur	Nerve to quadratus femoris	Laterally rotates thigh

7. Membrane
   Obturator
8. Muscles
   Erector spinae
   Erector spinae, iliocostalis part
   Erector spinae, longissimus part
   Erector spinae, spinalis part
   External oblique of abdomen
   Gemellus inferior
   Gemellus superior
   Gluteus maximus
   Gluteus medius
   Gluteus minimus
   Internal oblique of abdomen
   Latissimus dorsi
   Obturator internus
   Piriformis
   Posterior inferior serratus
   Quadratus femoris
9. Nerves
   Inferior gluteal
   Posterior femoral cutaneous
   Pudendal
   Sciatic
   Superior gluteal
10. Surface anatomy
   Anus
   Gluteal region
   Gluteal sulcus
   Lumbar region
   Lumbar triangle
   Median sulcus
11. Veins
   Inferior gluteal
   Internal pudendal
   Superior gluteal

SPINAL CORD, MENINGES, AND SACRAL PLEXUS

Dissection. Expose (p. 342) the entire posterior surface of the vertebral column and sacrum by removing all the muscles. Expose the sacrococcygeal joint from behind, and separate the coccyx from the sacrum by a horizontal section. Remove the piriformis muscle completely from the gluteal region. Remove the sacrotuberous and sacrospinous ligaments from the sacrum. With a chisel remove the posterior vertebral arches from T1 down to the last sacral vertebra and open the vertebral canal.

Study the posterior aspect of the dura mater (p. 344) and observe the epidural space. Cut the dura mater longitudinally, following the median plane. Observe and subsequently cut the arachnoid mater in a similar manner (p. 346).

Study the posterior aspect of the spinal cord, which is still covered with the pia mater, and in the upper part expose several projections of the denticulate ligament. Note the cervical and lumbar enlargements of the spinal cord, the conus medullaris, the filum terminale, the cauda equina, the ventral and dorsal roots of the spinal nerves, and the spinal ganglia.

Dissect the following nerves from the gluteal region toward the spinal cord, while removing piecemeal the sacral structure: sciatic, posterior femoral cutaneous, superior and inferior gluteal, and pudendal.

Study the composition of the spinal nerves and the brachial, lumbar, and sacral plexuses.

Cut the spinal nerves close to their ganglia and remove the spinal cord, together with the roots of the spinal nerves, from the vertebral canal. Study the external and internal morphology of the spinal cord in horizontal sections.

Dissect and subsequently study the following structures within the thoracic segment of the vertebral column: the posterior longitudinal ligament, the costovertebral and costotransverse joints (p. 231), and the joints between the vertebral bodies (p. 232) and between the articular processes. Open some of the representative joints and observe the articular surfaces, the intervertebral discs, and the intraarticular morphology.

Summary of structures

1. Bones and bony landmarks
   Coccyx
   Sacrum
   Vertebral arches
   Vertebral canal
   Vertebral column

Table 31

Name	Origin	Insertion	Innervation	Action
Piriformis (see Table 30)				

2. Central nervous system
   Cauda equina
   Conus medullaris
   Filum terminale
   Spinal cord
   Spinal cord, cervical enlargement
   Spinal cord, lumbar enlargement
   Spinal ganglia
   Spinal nerve, dorsal root
   Spinal nerve, ventral root
3. Joints
   Articular processes
   Costotransverse
   Costovertebral
   Intervertebral disc
   Sacrococcygeal
   Vertebral bodies
4. Ligaments
   Posterior longitudinal
   Sacrospinous
   Sacrotuberous
5. Meninges
   Arachnoid mater
   Denticulate ligament
   Dura mater
   Epidural space
   Pia mater
6. Muscle
   Piriformis
7. Nerves
   Inferior gluteal
   Posterior femoral cutaneous
   Pudendal
   Sciatic
   Spinal
   Superior gluteal
8. Plexuses
   Cervical
   Lumbar
   Sacral

PELVIS: POSTERIOR APPROACH

Dissection. Cut (p. 348) the vertebral column horizontally in between the sacrum and the L5 vertebra. Remove the remains of the sacrum. Locate the posterior superior iliac spine and, 4 cm lateral to it, cut the hip bone vertically from the iliac crest down to the greater sciatic notch. Remove the separated parts of the skeleton. Expose the posterior aspect of the rectum down to just above the anus (p. 350), and on that aspect dissect the division of the inferior mesenteric artery into the superior rectal vessels. Follow the latter vessels along the lateral rectal wall.

Complete the dissection of the superior and inferior gluteal and the internal pudendal arteries from the gluteal region to the internal iliac artery. Remove the accompanying veins. Complete the dissection of the middle rectal arteries and note how they communicate with the superior vessels.

Reflect the rectum posteroinferiorly as far as possible.

MALE

Dissection. After the rectum has been reflected, observe the rectovesical septum, which covers the posterior surface of the urinary bladder and the prostate. Remove the septum gradually downward while following the ductus deferentes over the posterior aspect of the prostate, at which level they are closely related to the seminal vesicles (p. 352). Expose the posterior aspect of the seminal vesicles down to where they join with the ductus deferentes (p. 355) and expose the ejaculatory ducts to the points at which they enter into the prostate.

Introduce a sound through the urethral orifice in the urinary bladder (p. 354) for about 1 cm and thus determine the position of the urethral duct inside the prostate.

FEMALE

Dissection. After the rectum has been reflected (p. 356), observe the rectovaginal septum which covers the posterior surface of the vagina. Remove the septum gradually downward and expose the entire posterior surface of the uterus and the vagina. Reflect these organs backward to dissect their anterior surfaces.

Observe the uterovesical septum, which is located between the anterior aspect of the uterus and the posterior aspect of the urinary bladder. Remove the septum.

Introduce a sound through the urethral orifice in the urinary bladder as far as the external orifice. Dissect the inferior aspect of the urinary bladder all around the urethra, while preserving the pelvic diaphragm which supports the urinary bladder and the other pelvic viscera in situ.

NOTE. In the course of this dissection all students are advised to follow, in addition to their own dissection, the dissection of a cadaver of the opposite sex (pp. 358 and 359).

Summary of structures

1. Arteries
   Inferior gluteal
   Inferior mesenteric
   Internal iliac
   Internal pudendal
   Middle rectal
   Superior gluteal
   Superior rectal

2. Bones and bony landmarks
   Greater sciatic notch
   Hip
   Iliac crest
   L5 vertebra
   Posterior superior iliac spine
   Sacrum
   Vertebral column
3. Diaphragm
   Pelvic
4. Organs
   Anus
   Ductus deferens
   Ejaculatory duct
   Pelvic viscera
   Prostate gland
   Rectum
   Seminal vesicle
   Urethra
   Urethral orifice, external
   Urethral orifice, internal
   Urinary bladder
   Uterus
   Vagina
5. Septa
   Rectovaginal
   Rectovesical
   Uterovesical

Part Six
PERINEUM

MALE

Observation. Observe the limits of the region and the underlying bony parts (p. 363): coccyx, ischial tuberosity, ischiopubic rami, and pubic symphysis. In addition, observe from posterior to anterior the anus, scrotum with testes, penis, median raphe, prepuce, frenulum, glans, and external urethral orifice (p. 364).

Palpation. Palpate the spermatic cord through the skin, from the superficial inguinal ring down into the scrotum as far as the epididymis and testis.

Dissection. Incise the skin (Fig. 11) from the posterior aspect of the anus, around the anus, along the median raphe, to the level of the penile neck. Separate the testes from one another. Cut the penile skin circularly from the ventral aspect of the organ around the penile neck.

Dissect the spermatic cord from the inguinal canal as far as the epididymis. Observe the tunica vaginalis, the parietal sheath, on that part of the testicular surface which is not covered by the epididymis (p. 375). Cut through the tunica vaginalis along the anterior testicular border and observe the vaginal space and the line of reflection between the parietal and visceral layers.

Dissect the components of the spermatic cord. Remove

Fig. 11

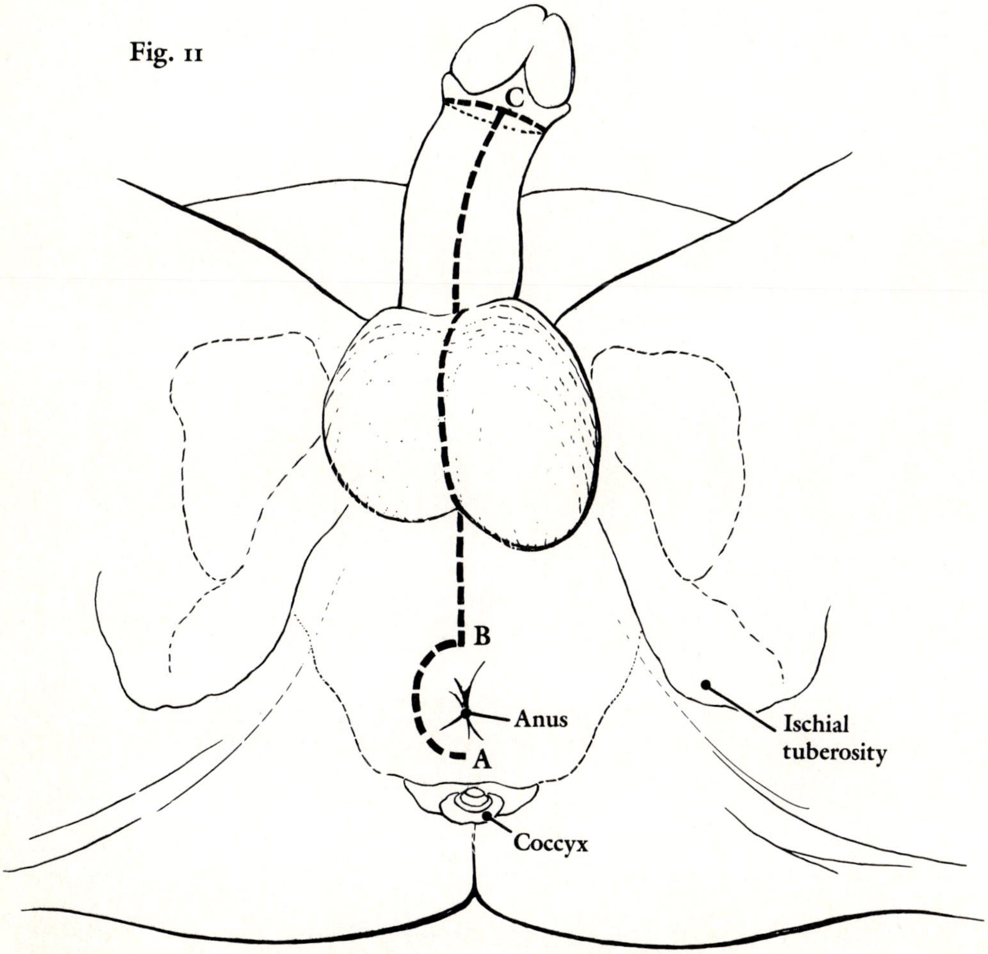

the spermatic cord, epididymis, and testes from the scrotum. Cut the testes, together with the epididymides, sagittally and study their interiors (p. 376).

Remove the skin from the entire surface of the perineum and penis.

Expose the sphincter ani externus muscle and observe its fibers attached to the coccyx posteriorly and to the perineal body, or the central tendon of the perineum anteriorly (p. 366). Isolate, in relation to the sphincter muscle, the inferior rectal artery, vein, and nerve from the internal pudendal vessels and the pudendal nerve. Dissect the rectal vessels and the nerve toward their origins while exposing the inferior aspect of the levator ani muscle from the level of the sphincter ani muscle as far as the bony attachments.

Follow the internal pudendal neurovascular bundle from the gluteal region, through the lesser sciatic notch anteriorly, toward the internal aspect of the corresponding ischiopubic ramus.

Expose the superficial transverse perineal muscle from the central tendon of the perineum to the ischial tuberosity.

Peel carefully the sphincter ani externus muscle from the rectal wall to expose the circular fibers of the sphincter ani internus muscle.

Dissect the bulbospongiosus muscle from the central tendon of the perineum anteriorly, and observe how the muscles from both sides join each other along the median plane to form the median raphe. These muscles, in their anterior parts, separate and curve around and subsequently attach to the lateral aspect of the penile root.

Expose the ischiocavernosus muscles from the inferior borders of the ischiopubic rami toward the penile root.

Expose the internal pudendal blood vessels and the pudendal nerve in the triangular space limited by the superficial transverse perineal muscle, bulbospongiosus muscle, and ischiocavernosus muscle.

FEMALE

Observation. Observe the limits of the region and the underlying bony parts (p. 363): coccyx, ischial tuberosity, ischiopubic rami, and pubic symphysis. In addition, observe from posterior to anterior the anus, labia majora, which surround the pudendal cleft, and posterior and anterior vaginal commissures. Reflect the labia majora laterally and observe the labia minora. Reflect the labia minora laterally and observe the vaginal vestibule, the external urethral orifice, the clitoris, the clitoral prepuce, and the clitoral frenulum (p. 365).

Palpation. Palpate the observed elements.

Dissection. Incise the skin (Fig. 12) from behind, from the posterior aspect of the anus forward, circling around the anus and the vaginal vestibule, in between the labia majora and minora, to the pubic symphysis. Remove the skin completely, together with the labia majora, from the perineum.

Expose the sphincter ani externus muscle and observe its

Fig. 12

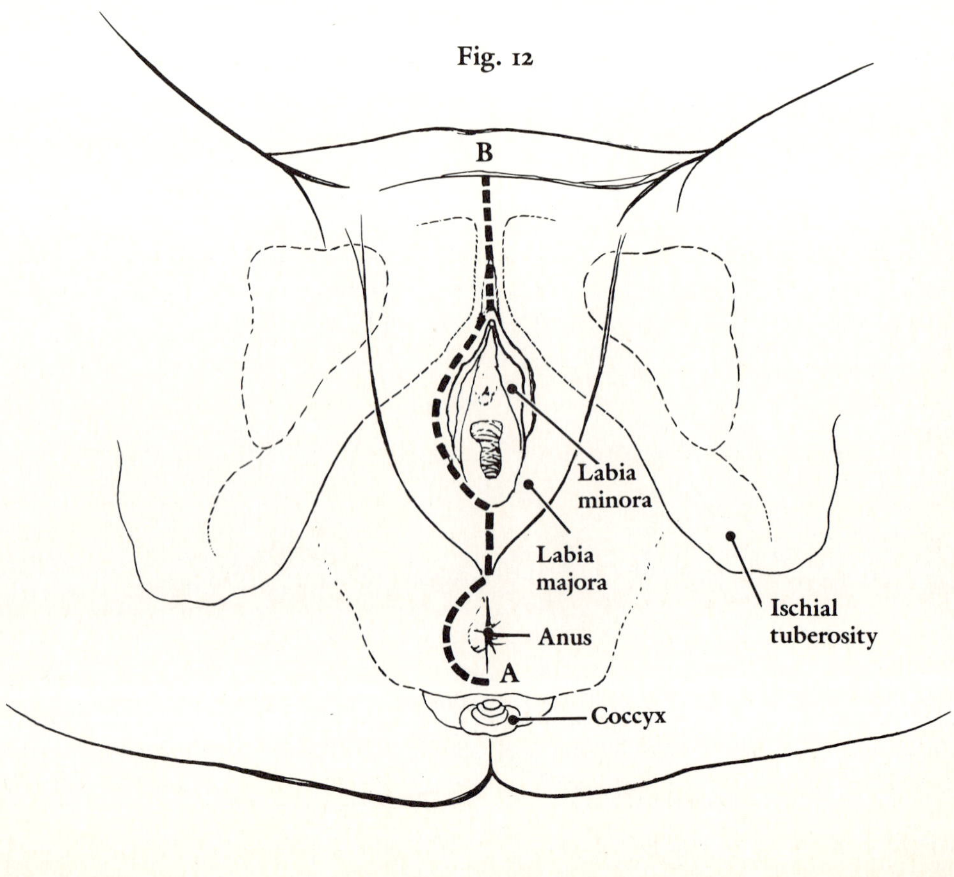

fibers attached to the coccyx posteriorly and to the central tendon of the perineum anteriorly (p. 368). Isolate, in relation to the sphincter muscle, the inferior rectal artery, vein, and nerve from the internal pudendal vessels and the pudendal nerve. Dissect the rectal vessels and the nerve toward their origins while exposing the inferior aspect of the levator ani muscle, from the level of the sphincter ani muscle as far as the bony attachments.

Follow the internal pudendal neurovascular bundle from the gluteal region, through the lesser sciatic notch anteriorly, toward the internal aspect of the corresponding ischiopubic ramus.

Expose the superficial transverse perineal muscle from the central tendon of the perineum to the ischial tuberosity.

Peel carefully the sphincter ani externus muscle from the rectal wall to expose the circular fibers of the sphincter ani internus muscle.

Locate by palpation, starting from the clitoral glans, the direction of the crura of the clitoris. They extend laterally and posteriorly to the ischiopubic rami.

Expose the bulbospongiosus muscles from the central tendon of the perineum anteriorly. They extend along the corresponding side of the vaginal vestibule to terminate on the clitoral crura.

Dissect the ischiocavernosus muscles over the entire extent of the clitoral crura, starting from the ischiopubic rami.

Expose the internal pudendal blood vessels and the pudendal nerve in the triangular space limited by the superficial transverse perineal muscle, bulbospongiosus muscle, and the ischiocavernosus muscle.

NOTE. In the course of this dissection all students are advised to follow, in addition to their own dissection, the dissection of a cadaver of the opposite sex.

Summary of structures

1. Arteries
   Inferior rectal
   Internal pudendal
2. Bones and bony landmarks
   Coccyx
   Ischial tuberosity
   Ischiopubic ramus
   Lesser sciatic notch
   Pubic symphysis
3. Canal
   Inguinal
4. Muscles
   Bulbospongiosus
   Ischiocavernosus
   Levator ani
   Sphincter ani externus
   Sphincter ani internus
   Superficial transverse perineal
5. Nerves
   Inferior rectal
   Pudendal
6. Organs
   Clitoral crura
   Epididymis
   Penile crura
   Root of penis
   Spermatic cord
   Testes
   Tunica vaginalis, parietal sheath
   Tunica vaginalis, visceral sheath
   Tunica vaginalis testis

Table 32

Name	Origin	Insertion	Innervation	Action
Bulbospongiosus	Median raphe and central tendon of perineum	Corpus cavernosum of penis	Perineal branch of pudendal nerve	Empties urethra; erection; sphincter of vagina
Ischiocavernosus	Inner surface of ischial tuberosity and ramus of ischium	Crus of penis	Perineal branch of pudendal nerve	Erection
Levator ani (see Table 29)				
Sphincter ani externus	Central tendon of perineum	Anococcygeal ligament	Perineal branch of S4 spinal nerve and inferior rectal branch of pudendal nerve	Sphincter of anus
Sphincter ani internus			Sympathetic	Sphincter of anus
Superficial transverse perineal	Medial surface of ischial tuberosity	Central tendon of perineum	Perineal branch of pudendal nerve	Stabilizes central tendon of perineum

7. Surface anatomy
   - Anterior vaginal commissure
   - Anus
   - Central tendon of the perineum (perineal body)
   - Clitoris
   - External urethral orifice
   - Frenulum of clitoris
   - Frenulum of penis
   - Glans clitoris
   - Glans penis
   - Labia majora
   - Labia minora
   - Median raphe
   - Penile neck
   - Penis
   - Perineum
   - Posterior vaginal commissure
   - Prepuce of clitoris
   - Prepuce of penis
   - Pudendal cleft
   - Scrotum
   - Superficial inguinal ring
   - Vaginal vestibule
8. Veins
   - Inferior rectal
   - Internal pudendal

MALE (continued)

Dissection. Remove the bulbospongiosus and the ischiocavernosus muscles (p. 370) and demonstrate the corpus spongiosum and the corpora cavernosa. Trace these three erectile bodies all the way through the free part of the penis and observe how the glans of the corpus spongiosum covers the distal ends of the corpora cavernosa.

Dissect the superficial and deep dorsal veins on both sides of the deep fascia, on the dorsal aspect of the penis, and between the corpora cavernosa (p. 374). Follow the deep dorsal vein of the penis posteriorly, beneath the pubic symphysis, as it joins the prostatic plexus. Immediately lateral to the deep vein, still on the dorsal aspect of the penis, find the dorsal arteries and nerves and dissect them posteriorly toward the internal pudendal artery and the pudendal nerve. Remove the blood vessels and the nerves from the penis.

Separate the two corpora cavernosa from each other and from the corpus spongiosus, starting from the dorsal aspect of the penis. Free the anterior parts of the corpora cavernosa from the glans and reflect them toward the corresponding ischiopubic ramus. Reflect the corpus spongiosus inferoposteriorly. Cut with scissors the dorsal aspect of the glans to open the urethra, and continue exposing the urethra posteriorly along the bulb of the corpus spongiosus.

Cut sagittally through the pubic symphysis with a large knife or a saw.

With a knife cut the contents of the pelvis and the perineum sagitally, through the median plane, from anterior posteriorly.

Separate the two lower limbs and the two perineal and pelvic halves. Study the structures on the section (p. 358).

Cut the semirectum horizontally, about 3 cm above the anus, and remove the upper part. On the same side remove the halves of the urinary bladder, prostate, urethra, seminal vesicle, and the ductus deferens.

Dissect the entire superior aspect of the levator ani muscle. Remove this muscle from the bony attachments.

Observe the thick fascia that covers the obturator internus muscle. Observe the internal pudendal neurovascular bundle, which is contained in the fascia of the obturator internus muscle. Dissect the pelvic part of the obturator internus muscle and remove it subsequently.

Expose the obturator membrane within the obturator foramen.

FEMALE (continued)

Dissection. Remove the bulbospongiosus and ischiocavernosus muscles (p. 372).

Expose the corpus, bifurcation, and crura of the clitoris starting from the free end—the glans of the clitoris.

Observe the relationship of the corpora spongiosa with the vaginal vestibule.

Table 33

Name	Origin	Insertion	Innervation	Action
Bulbospongiosus (see Table 32)				
Ischiocavernosus (see Table 32)				
Levator ani (see Table 29)				
Obturator internus (see Table 30)				

Complete the dissection of the pudendal neurovascular bundle anteriorly.

Cut the pubic symphysis sagittally with a large knife or a saw. With a knife cut the contents of the pelvis and the perineum sagittally, through the median plane, from anterior posteriorly.

Separate the two lower limbs and the two perineal and pelvic halves. Study the structures on the section (p. 359).

Cut the semirectum horizontally, about 3 cm above the anus, and remove the upper part. On the same side remove the halves of the urinary bladder, urethra, uterus, vagina, uterine tube, and the ovary.

Dissect the entire superior aspect of the levator ani muscle. Remove this muscle from the bony attachments.

Observe the thick fascia that covers the obturator internus muscle. Observe the pudendal neurovascular bundle, which is contained in the fascia of the obturator internus muscle. Dissect the pelvic part of the obturator internus muscle and remove it subsequently.

Expose the obturator membrane within the obturator foramen.

NOTE. In the course of this dissection all students are advised to follow, in addition to their own dissection, the dissection of a cadaver of the opposite sex.

If the frontal sections of this region are available (pp. 378 and 379), study the disposition and relationships of the organs and the diaphragms.

Summary of structures

1. Arteries
   Dorsal, of penis
   Internal pudendal
2. Bones and bony landmarks
   Ischiopubic ramus
   Obturator foramen
   Pubic symphysis
3. Ducts
   Deferens
   Urethra
4. Erectile bodies
   Clitoral bifurcation
   Clitoral corpus
   Clitoral crura
   Clitoral glans
   Corpus cavernosum
   Corpus spongiosum
   Glans penis
5. Fasciae
   Deep, of penis
   Obturator internus
   Obturator membrane
6. Muscles
   Bulbospongiosus
   Ischiocavernosus
   Levator ani
   Obturator internus
7. Nerves
   Dorsal, of penis
   Pudendal
8. Organs
   Anus
   Ovary
   Penis
   Prostate
   Rectum
   Seminal vesicle
   Urinary bladder
   Uterine tube
   Uterus
   Vaginal vestibule
9. Plexus
   Prostatic venous
10. Veins
   Deep dorsal, of penis
   Internal pudendal
   Superficial dorsal, of penis

Part Seven
LOWER LIMB

FEMORAL TRIANGLE

Observation. Observe the following structures on the anteromedial aspect of the thigh: the quadriceps femoris and adductor musculatures, the great saphenous vein, the patella, and the ligamentum patellae (pp. 382 and 389).

Palpation. Palpate the elements along the anterior margin of the hip bone, starting from the anterior superior iliac spine and proceeding to the pubic symphysis. Palpate the following: lateral and inferior to the anterior superior iliac spine, the greater trochanter; from the pubic symphysis to the ischial tuberosity, the ischiopubic ramus; lateral and medial to the patella, the femoral condyles; the adductor tubercle on the medial femoral condyle (p. 384); the quadriceps femoris and the adductor muscles; and below the medial half of the inguinal ligament within the femoral triangle, the superficial inguinal lymph nodes.

Dissection. Incise the skin vertically (Fig. 13) from the midpoint of the inguinal ligament inferiorly to a point 2 cm below the apex of the patella. Cut the skin horizontally 2 cm below the patellar apex and reflect the flaps medially and laterally.

Complete the dissection of the cutaneous branches of the lumbar plexus in the subcutaneous tissue. These nerves include the lateral femoral cutaneous nerve and the femoral branch of the genitofemoral nerve.

Isolate in the subcutaneous tissue (p. 390) the great saphenous vein on the medial aspect of the thigh, and dissect the vein as far inferiorly as possible. Subsequently follow the vein superiorly as it pierces the deep fascia, and observe its tributaries—the external pudendal, superficial epigastric, and superficial circumflex iliac veins. Expose the

Table 34

Name	Origin	Insertion	Innervation	Action
Adductor longus	Anterior surface of pubis	Linea aspera and middle third of femur	Anterior division of obturator nerve	Adducts thigh
Iliopsoas (see Table 28)				
Pectineus	Pecten pubis	From lesser trochanter to linea aspera	Femoral nerve	Adducts and flexes thigh
Quadriceps femoris	Anterior inferior iliac spine and capsule of hip joint (rectus); intertrochanteric line, greater trochanter, and linea aspera (vastus lateralis); intertrochanteric line, linea aspera, and medial intermuscular septum (vastus medialis); femoral shaft and lateral intermuscular septum (vastus intermedius)	Patella and tibial tuberosity	Femoral nerve	Extends leg; flexes thigh
Sartorius	Anterior superior iliac spine	Upper part of medial surface of tibia	Femoral nerve	Flexes thigh and leg; abducts and laterally rotates thigh

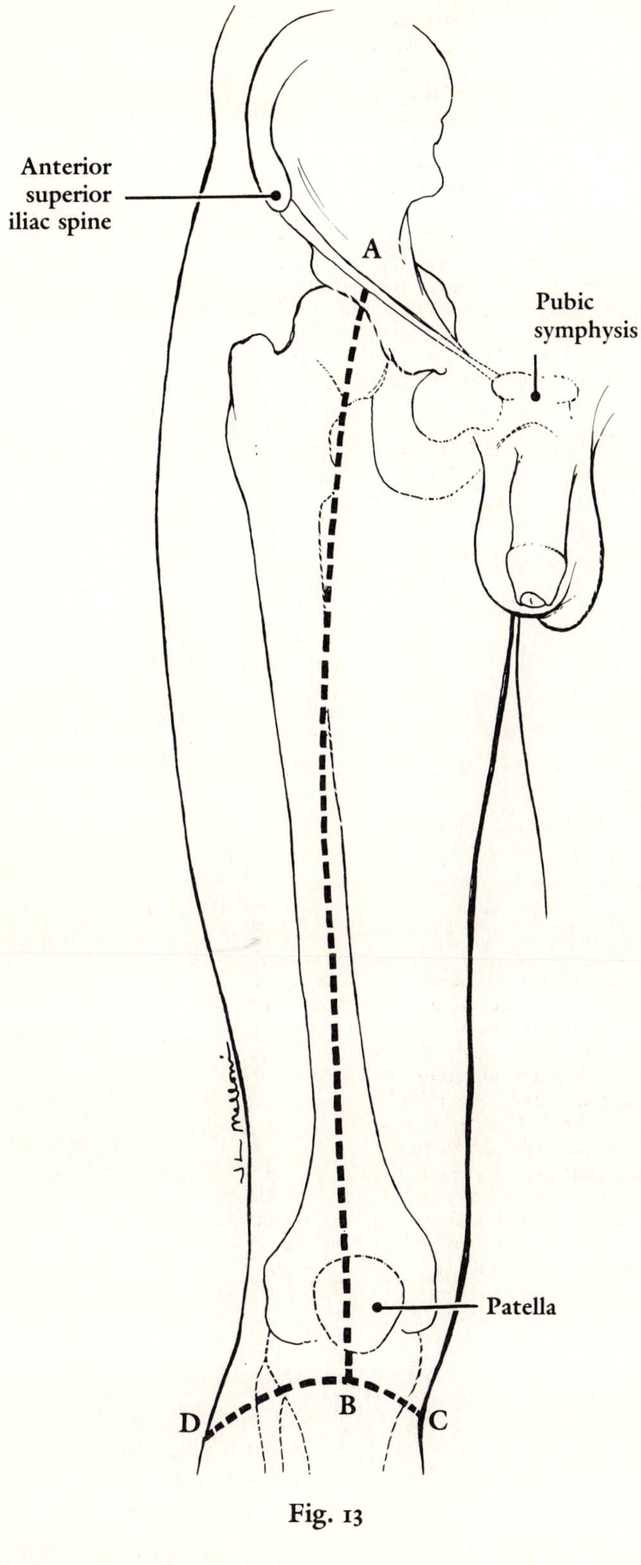
Anterior superior iliac spine
A
Pubic symphysis
Patella
D
B
C

Fig. 13

opening of the saphenous vein in the deep fascia and a number of superficial inguinal lymph nodes around this opening.

Clean the deep fascia over the femoral triangle and observe the branches of the femoral artery (external pudendal, superficial epigastric, superficial circumflex iliac) and the anterior cutaneous branches of the femoral nerve that pass through the fascia on their way from the deep into the subcutaneous layer. Clean the entire anterior and medial surfaces of the deep fascia, the fascia lata.

Cut the deep fascia from the anterior superior iliac spine down to the medial side of the knee joint, along the sartorius muscle. Expose the muscle by removing the fascia from its anterior aspect, and identify the blood vessels and nerves that enter the muscle through its medial border. Reflect the medial flap of the deep fascia medially (p. 392).

Dissect the most anterior adductor muscle, the adductor longus, from the pubic region obliquely down to the femur.

Observe the limits of the femoral triangle: the inguinal ligament superiorly, the sartorius muscle laterally, and the adductor longus muscle medially. Dissect, within the triangular space, the femoral artery and the following arterial branches: external pudendal, profunda femoris, medial circumflex, and lateral circumflex. Expose the femoral vein, which is in a medial position and parallel to the artery, and the junction of this vein with the great saphenous vein. Remove the other tributaries of the femoral vein. Demonstrate, immediately medial and parallel to the vein, the femoral canal, the fatty tissue, and the deep inguinal lymph nodes, which are contained in the canal. Introduce a finger from the femoral canal into the abdominal cavity by way of the femoral ring. Review the boundaries of the femoral ring. Immediately lateral and parallel to the femoral artery is the femoral nerve. Differentiate the terminal branches of this nerve within the triangle. Keep the contents of the femoral triangle intact while exposing the floor of this region, which is formed by two muscles—the iliopsoas muscle laterally and the pectineus muscle medially.

Summary of structures

1. Arteries
   - External pudendal
   - Femoral
   - Lateral circumflex
   - Medial circumflex
   - Profunda femoris
   - Superficial circumflex iliac
   - Superficial epigastric
2. Bones and bony landmarks
   - Adductor tubercle
   - Anterior superior iliac spine
   - Apex of patella
   - Femur
   - Greater trochanter
   - Hip
   - Ischial tuberosity
   - Ischiopubic ramus
   - Lateral femoral condyle
   - Medial femoral condyle
   - Patella
   - Pubic symphysis
3. Fasciae
   - Deep
   - Lata
4. Joint
   - Knee
5. Ligaments
   - Inguinal
   - Patellar
6. Lymph nodes
   - Deep inguinal
   - Superficial inguinal
7. Muscles
   - Adductor longus
   - Adductors
   - Iliopsoas
   - Pectineus
   - Quadriceps femoris
   - Sartorius
8. Nerves
   - Anterior femoral cutaneous
   - Femoral
   - Genitofemoral, femoral branch
   - Lateral femoral cutaneous
9. Plexus
   - Lumbar
10. Regions
    - Abdominal cavity
    - Anterior, of thigh
    - Femoral canal
    - Femoral ring
    - Femoral triangle
    - Medial, of thigh
    - Pubic
11. Veins
    - Femoral
    - Great saphenous
    - Superficial circumflex iliac
    - Superficial epigastric
    - Superficial external pudendal

ANTERIOR AND ADDUCTOR REGIONS OF THE THIGH

Dissection. Expose the tensor fasciae latae muscle from the anterior superior iliac spine and the iliac crest to its attachment on the deep surface of the fascia lata (p. 392). From this attachment down to the lateral aspect of the tibia, separate the iliotibial tract from the rest of the fascia lata by two vertical incisions. Remove that part of the fascia lata which covers the tensor muscle.

Dissect the rectus femoris muscle from the anterior inferior iliac spine down to the patella. Isolate the blood vessels and nerves that enter the muscle from the medial side.

Expose the anterior aspect of the vastus medialis and lateralis muscles, medial and lateral to the rectus muscle and on a deeper plane, down to the patella.

Reflect the sartorius muscle laterally and observe the strong fascia of the adductor canal (p. 394) deep to it, extending from the vastus medialis muscle to the adductor longus and magnus muscles.

Reflect the rectus femoris muscle and expose, deep to it, the vastus intermedius muscle, which is situated between the vastus medialis and lateralis muscles. While dissecting the heads of the quadriceps femoris muscle, identify simultaneously their blood vessels and nerves.

Complete the dissection of the muscular branches of the femoral nerve to the already exposed muscles.

Gradually cut the thick fascia between the vastus medialis muscle and the adductor muscles while following the femoral blood vessels downward and isolate the saphenous nerve, a branch of the femoral nerve, which joins the femoral vessels in the adductor canal. The nerve then pierces the fascia of the adductor canal to meet the great saphenous

Table 35

Name	Origin	Insertion	Innervation	Action
Adductor brevis	External surface of inferior ramus of pubis	Line from lesser trochanter to linea aspera and linea aspera	Obturator nerve	Adducts thigh
Adductor longus (see Table 34)				
Adductor magnus	Inferior ramus of pubis, ramus of ischium, and ischial tuberosity	Gluteal tuberosity of femur, linea aspera, medial supracondylar line of femur, and adductor tubercle of femur	Obturator nerve and tibial division of sciatic nerve	Adducts thigh
Gracilis	Medial margin of pubic body, inferior ramus of pubis, and ramus of ischium	Upper part of medial surface of tibia	Obturator nerve	Adducts thigh; flexes and medially rotates leg
Obturator externus	Rami of pubis, ramus of ischium, and external surface of obturator membrane	Trochanteric fossa of femur	Posterior division of obturator nerve	Laterally rotates thigh
Pectineus (see Table 34)				
Rectus femoris–quadriceps femoris (see Table 34)				
Sartorius (see Table 34)				
Tensor fasciae latae	Iliac crest, anterior superior iliac spine, and deep surface of fascia lata	Iliotibial tract	Superior gluteal nerve	Extends and laterally rotates leg; tenses fascia lata
Vastus intermedius–quadriceps femoris (see Table 34)				
Vastus lateralis–quadriceps femoris (see Table 34)				
Vastus medialis–quadriceps femoris (see Table 34)				

vein in the subcutaneous layer. Complete the dissection of the lateral circumflex artery.

Dissect the gracilis muscle from the pubic region to the knee joint, along the medial aspect of the thigh.

Cut the adductor longus and the pectineus muscles horizontally through the middle and reflect their parts upward and downward.

Find the obturator neurovascular bundle as it passes through the obturator membrane and dissect it deep to the already reflected pectineus and adductor longus muscles. Observe how the bundle divides into the anterior and posterior elements at the superior border of the adductor brevis muscle. Dissect the anterior obturator artery, vein, and nerve on the surface of the adductor brevis muscle. Next cut this muscle horizontally through the middle and reflect its parts upward and downward. Complete the dissection of the posterior obturator artery, vein, and nerve immediately deep to the already reflected adductor brevis muscle (p. 398).

Expose the anterior surface of the adductor magnus muscle and observe how the femoral vessels pierce through it from the anterior to the posterior compartment of the thigh. Complete the dissection of the medial circumflex and the profunda femoris arteries. Separate the perforating branches of the profunda femoris artery, which also pierce the adductor magnus muscle to enter into the posterior compartment of the thigh.

Dissect the obturator externus muscle from the lateral surface of the obturator membrane and the surrounding bony margins of the obturator foramen as far as possible. Complete the dissection of this muscle in the gluteal region.

Summary of structures

1. Arteries
   Femoral
   Lateral circumflex
   Medial circumflex
   Obturator
   Obturator, anterior branch
   Obturator, posterior branch
   Perforating branches of profunda femoris
   Profunda femoris
2. Bones and bony landmarks
   Anterior inferior iliac spine
   Anterior superior iliac spine
   Iliac crest
   Obturator foramen
   Patella
   Pubis
3. Canal
   Adductor
4. Fasciae
   Adductor canal
   Iliotibial tract
   Lata
5. Joint
   Knee
6. Membrane
   Obturator
7. Muscles
   Adductor brevis
   Adductor longus
   Adductor magnus
   Gracilis
   Obturator externus
   Pectineus
   Rectus femoris
   Sartorius
   Tensor fasciae latae
   Vastus intermedius
   Vastus lateralis
   Vastus medialis
8. Nerves
   Femoral
   Femoral, muscular branches
   Obturator
   Obturator, anterior branch
   Obturator, posterior branch
   Saphenous
9. Veins
   Femoral
   Great saphenous
   Obturator
   Obturator, anterior branch
   Obturator, posterior branch

POSTERIOR REGION OF THE THIGH

Observation. Observe (p. 401) the hamstring musculature and, in the lower part of the region, two groups of muscular tendons, the one lateral and the other medial. These two tendinous groups diverge to the corresponding side of the knee joint to delimit a visible depression, especially with the leg flexed—the popliteal fossa.

Palpation. Palpate the hamstring musculature, the lateral and medial tendinous groups, and the skeletal elements of the thigh and the popliteal fossa (pp. 383 and 385).

Dissection. Incise the skin vertically (Fig. 14) from the midpoint of the gluteal sulcus down to the horizontal level about 2 cm below the patellar apex. Remove the skin flaps from the posterior aspect of the thigh.

Complete the dissection of the posterior femoral cutaneous nerve in the subcutaneous layer (p. 402).

Dissect the semitendinosus muscle from the ischial tuberosity, along the medial side of the region, down to the medial side of the knee joint. Reflect this muscle, and

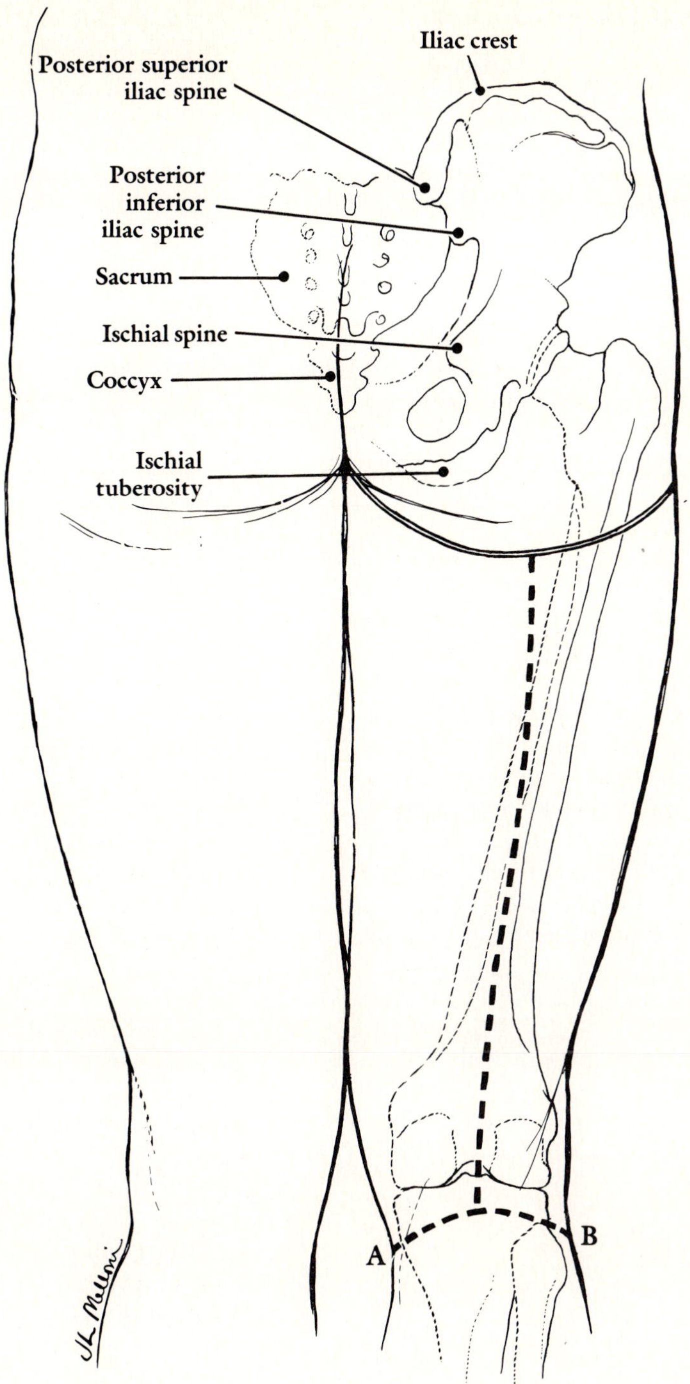

Fig. 14

deep to it expose the semimembranous muscle from the same tuberosity down to the medial side of the knee joint.

Dissect the long head of the biceps femoris muscle from the ischial tuberosity down to the head of the fibula. Reflect this head medially or laterally to expose the deeper, short head of the biceps femoris muscle. Observe how the two heads fuse before they become inserted into the fibular head.

Follow the sciatic nerve from the gluteal region as it passes in front of the long head of the biceps femoris muscle, between the lateral (two heads of the biceps femoris muscle) and medial (semitendinosus and semimembranosus muscles) muscles of the region, and finally to its bifurcation into the tibial and the common peroneal nerves. The level at which the nerve bifurcates varies considerably from one individual to another. Isolate the branches of the sciatic nerve to the hamstring muscles.

Table 36

Name	Origin	Insertion	Innervation	Action
Adductor magnus (see Table 35)				
Biceps femoris	Ischial tuberosity and sacrotuberous ligament (long head); lateral lip of linea aspera and lateral intermuscular septum (short head)	Head of fibula	Tibial division (long head) and common peroneal division (short head) of sciatic nerve	Flexes leg; laterally rotates thigh and leg
Semimembranosus	Ischial tuberosity	Medial surface of tibial condyle	Tibial division of sciatic nerve	Flexes leg; medially rotates thigh and leg
Semitendinosus	Ischial tuberosity	Medial surface of upper part of tibia	Tibial division of sciatic nerve	Flexes leg; medially rotates thigh and leg
Vasti (see Table 35)				

Reflect the medial muscular group medially and the lateral group laterally (p. 404).

Expose the posterior aspect of the adductor magnus muscle and the perforating branches of the profunda femoris artery in the posterior compartment of the thigh. Observe the connections between the neighboring perforating branches. Expose the opening of the femoral vessels in the adductor magnus muscle and from there follow the vessels into the popliteal fossa.

Study the attachments of the adductor and vastus muscles along the linea aspera of the femur.

Summary of structures

1. Arteries
   Femoral
   Perforating branches of profunda femoris
2. Bones and bony landmarks
   Fibula, head
   Ischial tuberosity
   Linea aspera
   Patellar apex
3. Fossa
   Popliteal
4. Joints
   Knee
5. Muscles
   Adductor magnus
   Adductors
   Biceps femoris, long head
   Biceps femoris, short head
   Hamstrings
   Semimembranosus
   Semitendinosus
   Vasti
6. Nerves
   Common peroneal
   Posterior femoral cutaneous
   Sciatic
   Sciatic, muscular branches
   Tibial
7. Regions
   Gluteal
   Posterior femoral
8. Surface anatomy
   Gluteal sulcus
9. Veins
   Femoral

POPLITEAL FOSSA AND POSTERIOR REGION OF THE LEG

Observation. Observe (p. 406) the musculature of the back of the leg, calcaneal tendon, tuber of the calcaneus, medial malleolus, lateral malleolus, and the small saphenous vein, which is sometimes visible along the subcutaneous layer of this aspect of the leg.

Palpation. In addition to the observed structures, palpate (pp. 383, 385, and 387) the upper extremity of the tibia, the head and neck of the fibula, and, around the lateral aspect of the fibular neck, the common peroneal nerve.

Dissection. Incise the skin (Fig. 15), following the midline of the region from the upper horizontal skin incision down to the tuber of the calcaneus. Cut the skin laterally

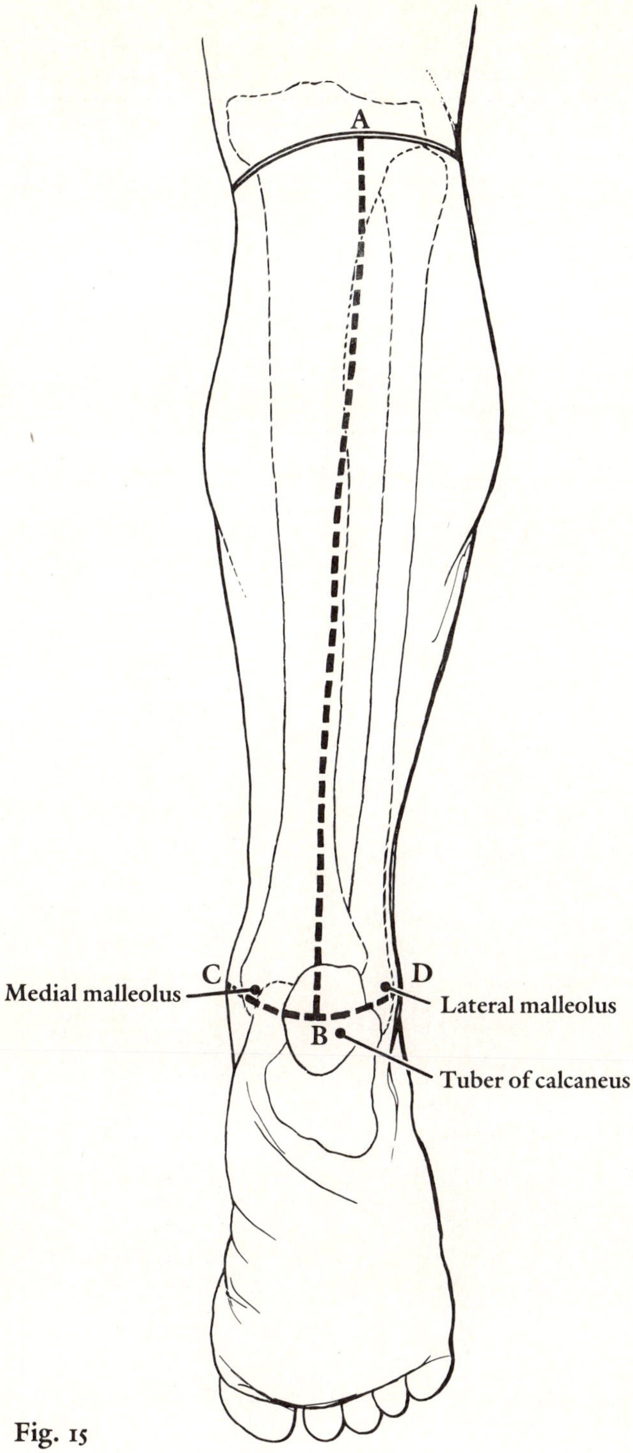

Fig. 15

and medially from the tuber of the calcaneus up to the tips of the lateral and medial malleoli. Reflect the skin flaps laterally and medially (p. 408).

Dissect along the posterior surface of the leg, starting from the posterior aspect of the lateral malleolus, the small saphenous vein up to its junction with the popliteal vein. Identify the sural nerve along the small saphenous vein, dissect the nerve upward, and separate the two branches: the medial sural cutaneous from the tibial nerve and the peroneal communicating branch from the common peroneal nerve, which form the sural nerve. Dissect these branches up to their origins.

Note the flexor retinaculum of the deep fascia that extends from the medial malleolus to the calcaneal tendon

Table 37

Name	Origin	Insertion	Innervation	Action
Adductor magnus (see Table 35)				
Biceps femoris (see Table 36)				
Gastrocnemius	Medial and lateral condyles of femur and capsule of knee joint	Tendon calcaneus, calcaneus	Tibial nerve	Flexes foot (plantar) and leg
Semimembranosus (see Table 36)				
Semitendinosus (see Table 36)				

and then from the tendon to the lateral malleolus.

Expose the posterior surface of the gastrocnemius muscle (p. 410). Dissect the two heads of this muscle, which originate from the corresponding femoral condyles. Observe how the heads join one another and continue by exposing the muscle downward along the calcaneal tendon to the tuber of the calcaneus.

Study the boundaries of the popliteal fossa: the biceps femoris muscle superolaterally, the semitendinosus and semimembranosus muscles superomedially, the lateral head of the gastrocnemius muscle inferolaterally, and the medial head of the gastrocnemius muscle inferomedially.

Observe the fatty tissue and the occasional lymph nodes within the subcutaneous layer of the popliteal fossa.

Expose the tibial and the common peroneal nerves from the bifurcation of the sciatic nerve throughout the popliteal fossa.

Follow the popliteal blood vessels from their opening in the adductor magnus muscle downward throughout the popliteal fossa. Identify the arterial branches for the neighboring muscles and the knee joint. Study the relationships of the elements in the popliteal fossa and simultaneously expose the floor of the fossa.

Summary of structures

1. Arteries
   Popliteal
   Popliteal, articular branches
   Popliteal, muscular branches
2. Bones and bony landmarks
   Fibula, head
   Fibula, neck
   Lateral femoral condyle
   Lateral malleolus
   Medial femoral condyle
   Medial malleolus
   Tibia
   Tuber of the calcaneus
3. Fasciae
   Deep
   Flexor retinaculum
4. Fossae
   Popliteal
   Popliteal, boundaries
   Popliteal, floor
5. Lymph nodes
   Popliteal
6. Muscles
   Adductor magnus
   Biceps femoris
   Gastrocnemius
   Gastrocnemius, lateral head
   Gastrocnemius, medial head
   Semimembranosus
   Semitendinosus
7. Nerves
   Common peroneal
   Medial sural cutaneous of tibial
   Peroneal communicating of common peroneal
   Sciatic, bifurcation
   Sural
8. Tendon
   Calcaneal
9. Veins
   Popliteal
   Small saphenous

POSTERIOR AND LATERAL REGIONS OF THE LEG

Dissection. Cut horizontally through the heads of the gastrocnemius muscle and reflect them to their corresponding femoral condyles. Reflect the lower muscular part as far downward as possible (p. 412).

Dissect the plantaris muscle, if present, from the lateral femoral condyle medially and inferiorly to the calcaneal tendon.

Expose the posterior surface of the soleus muscle and observe its attachments to the tibia and the fibula. Between these two bony attachments is the arch of the soleus muscle. While dissecting the muscle downward, note how it fuses with the gastrocnemius muscle to form the calcaneal tendon.

Continue the dissection of the tibial nerve and the popliteal blood vessels downward, as they pass beneath the arch of the soleus muscle.

Remove the plantaris muscle from the femoral condyle and the soleus muscle from the tibia and fibula. Reflect these muscles, together with the gastrocnemius muscle, down onto the tuber of the calcaneus (p. 414).

Proceed by dissecting the tibial nerve and the popliteal artery and vein until the vessels divide. Follow the anterior tibial artery from the bifurcation of the popliteal artery, together with the two accompanying veins, as it passes through the interosseous membrane to enter the anterior compartment of the leg.

Dissect the posterior tibial artery from the bifurcation of the popliteal artery, together with the two accompanying

Table 38

Name	Origin	Insertion	Innervation	Action
Flexor digitorum longus	Posterior surface of tibia and fascia of tibialis posterior muscle	Plantar surface of base of distal phalanges	Tibial nerve	Flexes digits and foot (plantar)
Flexor hallucis longus	Posterior surface of fibula, interosseous membrane, intermuscular septum, and fascia of tibialis posterior muscle	Posterior surface of base of distal phalanx of big toe	Tibial nerve	Flexes big toe and foot (plantar)
Gastrocnemius (see Table 37)				
Peroneus brevis	Lower two thirds of lateral surface of fibula and intermuscular septum	Tubercle of base of fifth metatarsal bone	Superficial peroneal nerve	Everts foot
Peroneus longus	Upper two thirds of lateral surface of fibula and intermuscular septum	Lateral side of base of first metatarsal bone and medial cuneiform bone	Superficial peroneal nerve	Everts foot
Plantaris	Lateral supracondylar line of femur and oblique popliteal ligament	Calcaneus	Tibial nerve	Flexes foot (plantar) and leg
Popliteus	Lateral condyle of femur, arcuate popliteal ligament, and lateral meniscus	Upper posterior surface of tibia	Tibial nerve	Medially rotates leg; laterally rotates thigh; flexes leg
Soleus	Posterior surface of fibular head and upper part of shaft, soleal line of tibia, and fibrous band between tibia and fibula	Tendo calcaneus, calcaneus	Tibial nerve	Flexes foot (plantar)
Tibialis posterior	Posterior surface of interosseous membrane, tibia, fibula, and intermuscular septa	Tuberosity of navicular bone and inferior surface of medial cuneiform bone	Tibial nerve	Inverts and flexes (plantar) foot

veins, through the posterior compartment of the leg to the medial retromalleolar region. Identify the main branch of this artery, the peroneal. Observe also the branches of the tibial nerve for the muscles in the posterior compartment of the leg.

Expose the popliteus muscle on the posterior aspect of the knee joint, from the lateral femoral condyle down to the tibia.

Dissect, in the deep muscular layer of the posterior compartment of the leg, the flexor digitorum longus muscle, the tibialis posterior muscle, and the flexor hallucis longus muscle, in that order from medial to lateral. While following these three muscles downward to the posterior aspect of the medial malleolus, observe how the tendon of the tibialis posterior muscle crosses anterior and then medial to the tendon of the flexor digitorum longus muscle. On the posterior aspect of the medial malleolus incise longitudinally the synovial tendinous sheaths of these three muscles and expose the tendons. Study the relationships of the tendons, blood vessels, and the nerve behind the medial malleolus, within the medial retromalleolar region.

Dissect the peroneus longus muscle along the lateral aspect of the fibula from the fibular neck down. Note how the tendon of this muscle curves around the posterior aspect of the lateral malleolus.

Dissect, deep to the peroneus longus muscle, the peroneus brevis muscle. Note how the tendon of this muscle curves around the posterior aspect of the lateral malleolus.

Follow the common peroneal nerve from the fibular neck, through the substance of the peroneus longus muscle, to its bifurcation into the deep and superficial peroneal nerves. Review the relationship of the common peroneal nerve at the fibular neck, where the nerve is readily palpable in living subjects.

Summary of structures

1. Arteries
   Anterior tibial
   Peroneal
   Popliteal
   Posterior tibial
2. Bones and bony landmarks
   Fibula
   Fibula, neck
   Lateral femoral condyle
   Lateral malleolus
   Medial femoral condyle
   Medial malleolus
   Tibia
   Tuber of the calcaneus
3. Joint
   Knee
4. Membrane
   Interosseous
5. Muscles
   Flexor digitorum longus
   Flexor hallucis longus
   Gastrocnemius, heads
   Peroneus brevis
   Peroneus longus
   Plantaris
   Popliteus
   Soleus
   Soleus, arch
   Tibialis posterior
6. Nerves
   Common peroneal
   Deep peroneal
   Superficial peroneal
   Tibial
   Tibial, muscular branches
7. Synovial structures
   Tendinous sheaths
8. Tendon
   Calcaneal
9. Veins
   Anterior tibial
   Popliteal
   Posterior tibial

SOLE OF THE FOOT

Observation. Observe (p. 417) the relief of the sole of the foot, the longitudinal plantar arch, and the transverse plantar arch.

Palpation. Palpate (p. 387) the skeletal elements of the foot.

Dissection. Incise the skin (Fig. 16) of the region from the tuber of the calcaneus to the root of the second toe. Proceed by incising the skin medially and laterally, from the root of the second toe to the roots of the big and the fifth toes, respectively. Reflect the parts of the skin laterally and medially. Remove the skin from the plantar surfaces of the toes after it has been incised along the digital midlines.

Expose the plantar aponeurosis (p. 418) and the five bands, from the tuber of the calcaneus to the digital roots. Observe the transverse metatarsal ligament, which keeps together the five digital bands. Reflect gradually the bands and subsequently the aponeurosis backward onto the tuber of the calcaneus by cutting the septa along the lateral and medial borders of the aponeurosis.

Dissect the flexor digitorum brevis muscle (p. 420) and

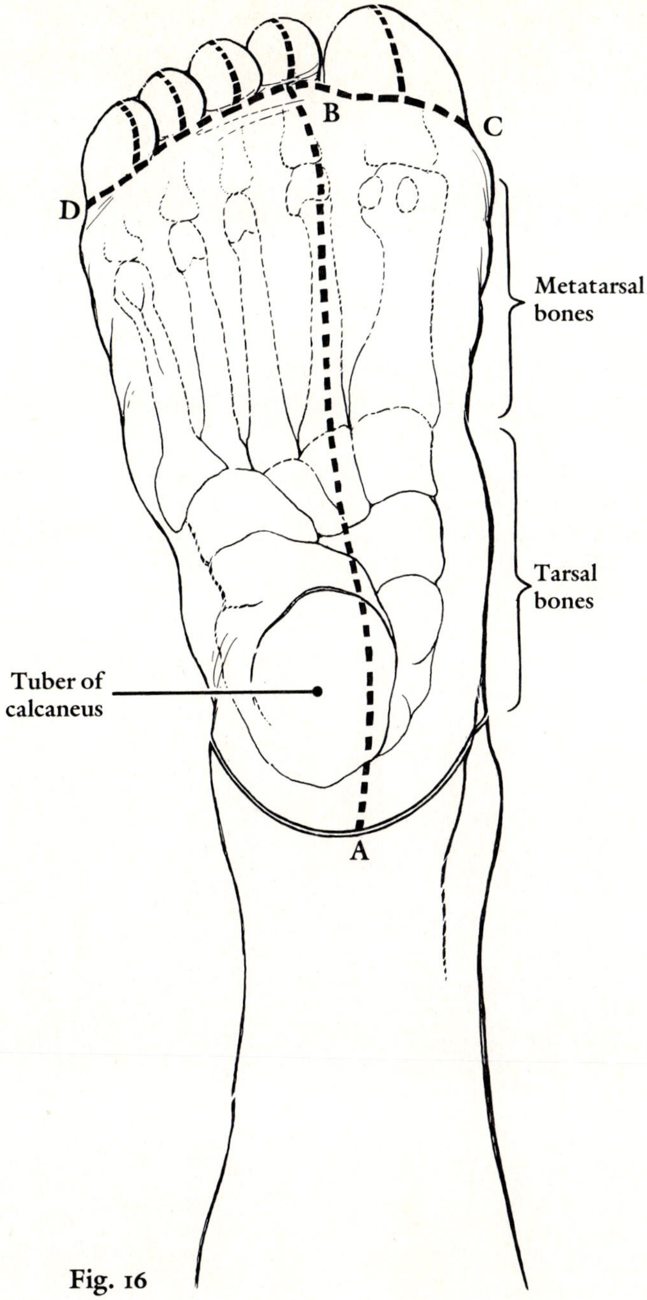

Fig. 16

its tendons from the tuber of the calcaneus to the corresponding digits. Cut the muscle in half across the middle and reflect the parts to the respective attachments. Expose, deep to the flexor digitorum brevis muscle, the lateral and medial plantar neurovascular bundles. Subsequently follow these nerves and blood vessels as far anteriorly as possible (p. 422).

Expose the abductor hallucis and the abductor digiti minimi muscles along the medial and lateral borders of the region, respectively.

Dissect the flexor accessorius muscle from the tuber of the calcaneus to the tendon of the flexor digitorum longus muscle. Expose simultaneously the tendon of the flexor digitorum longus muscle and its four terminal parts. Complete the dissection of the terminal tendons of the flexor digitorum longus muscle and, originating from them, of the four lumbrical muscles.

Complete the dissection of the tendon of the flexor hallucis longus muscle to the digital attachment. Next dissect, on both sides of this tendon, the two heads of the flexor hallucis brevis muscle.

Dissect the flexor digiti minimi brevis muscle from the base of the fifth metatarsal bone to the fifth digit.

Remove the flexor accessorius muscle from the tendon

Table 39

Name	Origin	Insertion	Innervation	Action
Abductor digiti minimi	Calcanean tuberosity, plantar aponeurosis, and intermuscular septum	Lateral side of base of proximal phalanx of fifth toe	Lateral plantar nerve	Abducts fifth toe
Abductor hallucis	Flexor retinaculum, calcanean tuberosity, plantar aponeurosis, and intermuscular septum	Medial side of base of proximal phalanx of big toe	Medial plantar nerve	Abducts big toe
Adductor hallucis	Bases of second, third, and fourth metatarsal bones, sheath of tendon of peroneus longus muscle, plantar metatarsophalangeal ligaments of third, fourth, and fifth toes, and deep transverse metatarsal ligament	Base of proximal phalanx of big toe	Deep branch of lateral plantar nerve	Adducts big toe
Flexor accessorius	Calcaneus	Tendon of flexor digitorum longus	Lateral plantar nerve	Flexes digits
Flexor digiti minimi	Base of fifth metatarsal bone and sheath of peroneus longus muscle	Lateral side of base of proximal phalanx of fifth toe	Superficial branch of lateral plantar nerve	Flexes fifth toe
Flexor digitorum brevis	Calcanean tuberosity, plantar aponeurosis, and intermuscular septa	Shafts of middle phalanges	Medial plantar nerve	Flexes digits
Flexor digitorum longus (see Table 38)				
Flexor hallucis brevis	Cuboid and lateral cuneiform bones, tendon of tibialis posterior muscle, and intermuscular septum	Medial and lateral sides of base of proximal phalanx of big toe	Medial plantar nerve	Flexes big toe
Flexor hallucis longus (see Table 38)				
Plantar interossei	Base of medial side of metatarsal bones	Medial side of base of proximal phalanx of same toe and dorsal digital expansion	Branches of lateral plantar nerve	Adduct third, fourth, and fifth toes
Lumbricals	Tendons of flexor digitorum longus muscle	Dorsal digital expansion	Medial plantar nerve (first) and deep branch of lateral plantar nerve (second, third, and fourth)	Flex metatarsophalangeal joints; extend interphalangeal joints

of the flexor digitorum longus muscle and reflect it backward onto the tuber of the calcaneus. Cut the tendon of the flexor digitorum longus muscle and reflect the four terminal tendons, together with the lumbrical muscles, onto the digits.

Dissect the transverse and oblique heads of the adductor hallucis muscle toward the base of the big toe and subsequently remove it (p. 424).

Complete the dissection of the lateral plantar artery, the plantar arch, to the first intermetatarsal space, where it anastomoses with the dorsalis pedis artery.

Expose the plantar interosseous muscles.

Summary of structures

1. Aponeuroses
   Lateral septum
   Medial septum
   Plantar
   Plantar, bands
   Transverse metatarsal ligaments
2. Arteries
   Dorsalis pedis
   Lateral plantar
   Medial plantar
   Plantar arch
3. Bones and bony landmarks
   Fifth metatarsal, base
   First intermetatarsal space
   Foot skeleton
   Tuber of the calcaneus
4. Muscles
   Abductor digiti minimi
   Abductor hallucis
   Adductor hallucis, oblique head
   Adductor hallucis, transverse head
   Flexor accessorius
   Flexor digiti minimi
   Flexor digitorum brevis
   Flexor digitorum longus, tendon
   Flexor hallucis brevis, heads
   Flexor hallucis longus, tendon
   Interosseous, plantar
   Lumbricals
5. Nerves
   Lateral plantar
   Medial plantar
6. Surface anatomy
   Digital roots
   Longitudinal plantar arch
   Plantar digital surfaces
   Transverse plantar arch
7. Veins
   Lateral plantar
   Medial plantar

ANTERIOR REGION OF THE LEG AND DORSUM OF THE FOOT

Observation. Observe (p. 426) the anterior border of the tibia, the medial and lateral malleoli, and the subcutaneous venous arch on the dorsum of the foot.

Palpation. Palpate (pp. 384 and 386) the upper extremities of the tibia and fibula, the anterior border and the medial surface of the tibia, the medial and lateral malleoli, and the skeleton of the foot.

Dissection. Remove the skin from the anterior aspect of the leg, the dorsum of the foot, and the dorsal aspect of the toes (Fig. 17). Complete the dissection of the superficial peroneal nerve in the subcutaneous tissue from where it pierces the deep fascia downward into the foot (p. 426).

Complete the dissection of the small saphenous vein and the sural nerve in the foot. Follow also the great saphenous vein, which is accompanied by the saphenous nerve, inferiorly into the foot. Expose the venous arch on the dorsal aspect of the foot.

Identify the thickened parts of the deep fascia—the superior and inferior extensor retinacula—and separate them from the rest of the fascia by section. Remove the entire deep fascia of the region, with the exception of the retinacula.

Dissect the tibialis anterior muscle and its tendon from the upper part of the lateral aspect of the tibia as far as the insertion into the foot, while preserving the retinacula.

Observe and subsequently dissect the extensor digitorum longus muscle, which is immediately lateral to the tibialis anterior muscle. First expose the muscle down to the superior retinaculum, then between the superior and inferior retinacula, and finally trace the five terminal tendons to their respective insertions. Although the four medial tendons terminate in the corresponding digits, the fifth tendon, of the peroneus tertius muscle, inserts into the base of the fifth metatarsal bone.

Reflect the extensor digitorum longus muscle laterally and the tibialis anterior muscle medially, and between them and on a deeper plane identify the extensor hallucis longus muscle. Follow this muscle and then the tendon of it beneath the extensor retinacula toward the big toe.

Find the anterior tibial blood vessels and the deep pe-

Fig. 17

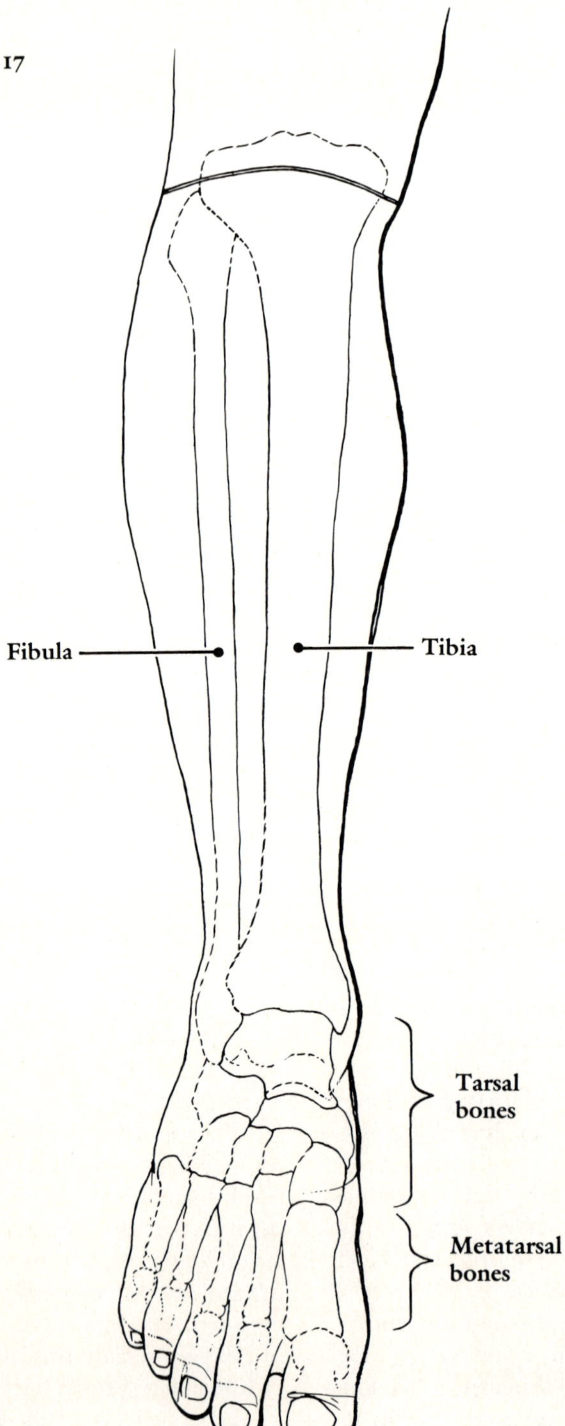

Table 40

Name	Origin	Insertion	Innervation	Action
Dorsal interossei	Adjacent sides of metatarsal bones	Bases of proximal phalanges and dorsal digital expansion	Branches of lateral plantar nerve	Stabilize second toe; abduct third and fourth toes
Extensor digitorum brevis	Calcaneus, interosseous talocalcaneal ligament, and inferior extensor retinaculum	Tendons of extensor digitorum longus muscle	Deep peroneal nerve	Extends second, third, and fourth toes
Extensor digitorum longus	Lateral condyle of tibia, medial surface of fibula, interosseous membrane, and intermuscular septum	Dorsal digital expansion	Deep peroneal nerve	Extends lateral four toes; dorsiflexes foot
Extensor hallucis brevis	Calcaneus, interosseous talocalcaneal ligament, and inferior extensor retinaculum	Dorsal surface of base of proximal phalanx of big toe	Deep peroneal nerve	Extends big toe
Extensor hallucis longus	Medial surface of fibula and interosseous membrane	Dorsal surface of base of distal phalanx of big toe	Deep peroneal nerve	Extends big toe; dorsiflexes foot
Peroneus tertius	Lower third of medial surface of fibula, interosseous membrane, and intermuscular septum	Dorsal surface of base of fifth metatarsal bone	Deep peroneal nerve	Dorsiflexes foot
Tibialis anterior	Lateral condyle and upper two thirds of tibia, interosseous membrane, and intermuscular septum	Talus, medial cuneiform bone, and head of first metatarsal bone	Deep peroneal nerve	Dorsiflexes and inverts foot

roneal nerve between the tibialis anterior and the extensor hallucis longus muscles (p. 428). Follow the nerve and the blood vessels upward, toward their respective origins. Keep the retinacula intact while dissecting the same neurovascular bundle from the leg into the foot, where the anterior tibial artery becomes continuous with the dorsalis pedis artery. Expose the latter artery until it pierces the contents of the first intermetatarsal space to anastomose with the plantar arch. Isolate the main branch of the dorsalis pedis artery, the arcuate artery.

Dissect the surfaces of the extensor hallucis brevis and the extensor digitorum brevis muscles on the dorsal aspect of the foot, from the tarsal sinus to the corresponding digits (p. 430).

Cut the extensor retinacula along the already dissected tendons and observe the extensor tendinous synovial sheaths.

Expose the dorsal interosseous muscles.

Complete the dissection of the dorsalis pedis artery in the first intermetatarsal space and find the anastomosis with the plantar arch.

Summary of structures

1. Arteries
   - Anterior tibial
   - Arcuate
   - Dorsalis pedis
   - Plantar arch
2. Bones and bony landmarks
   - Fibula, upper extremity
   - Fifth metatarsal, base
   - First intermetatarsal space
   - Foot skeleton
   - Lateral malleolus
   - Medial malleolus
   - Tarsal sinus
   - Tibia, anterior border
   - Tibia, lateral surface
   - Tibia, medial surface
   - Tibia, upper extremity

3. Fasciae
   Deep
   Inferior retinaculum
   Superior retinaculum
4. Muscles
   Dorsal interossei
   Extensor digitorum brevis
   Extensor digitorum longus
   Extensor hallucis brevis
   Extensor hallucis longus
   Peroneus tertius
   Tibialis anterior
5. Nerves
   Deep peroneal
   Saphenous
   Superficial peroneal
   Sural
6. Sheath
   Synovial, extensor tendinous
7. Veins
   Anterior tibial
   Great saphenous
   Small saphenous
   Subcutaneous dorsal arch

JOINTS OF THE LOWER LIMB

Hip joint. Reflect the iliopsoas muscle to the lesser trochanter and observe the bursa on the deep aspect of the muscle. Remove the muscle.

Cut the rectus femoris muscle through the middle and reflect the upper part superiorly. Observe the two tendons of this muscle. One is attached to the anterior inferior iliac spine and the other to the acetabular margin.

Remove all muscles from the greater trochanter and all the adductor muscles.

Dissect the articular capsule on the anterior aspect of the hip joint and expose the pubofemoral and iliofemoral ligaments (p. 432). Proceed by dissecting the superior and then the posterior surfaces of the articular capsule. Subsequently expose the ischiofemoral ligament on the posterior aspect of the joint (p. 433).

Cut the articular capsule circularly and reflect the parts to the hip bone and the femur. Demonstrate the ligament of the head of the femur by removing the femoral head from the acetabulum (p. 434). Cut the ligament of the head of the femur through the middle and separate the articular surfaces. Observe the transverse ligament in the acetabulum.

Knee joint. Remove the heads of the quadriceps femoris muscle from the femur and reflect them down onto the patella. Expose the articularis genu muscle. Observe the aponeurotic expansions of the vastus muscles, which extend from the patella and the ligamentum patellae laterally and medially (p. 435).

Complete the dissection of the ligamentum patellae and the adipose bursa, which is situated deep to the ligament.

Remove the following muscles from the knee region: sartorius, adductor magnus, gracilis, semitendinosus, biceps femoris, gastrocnemius, plantaris, and popliteus.

Dissect the fibular and the tibial collateral ligaments on the lateral and medial aspects of the joint, respectively.

Expose the oblique popliteal ligament on the posterior aspect of the joint, and complete the dissection of the semimembranosus muscle down to the medial aspect of the tibia (p. 437).

Observe the attachments of the articular capsule to the femur, patella, and the tibia.

Reflect the patella, together with the quadriceps femoris muscle, downward while at the same time removing the articular capsule from the anterior aspect of the joint. Once the capsule is completely removed from the patella, reflect the ligamentum patellae onto the tibial tuberosity. Incise the articular capsule and the ligaments horizontally, around the joint, and reflect the parts to their respective attachments.

Demonstrate from in front, with the leg flexed, the anterior and posterior cruciate ligaments inside the knee joint (p. 436). Cut these ligaments to separate the articular surfaces.

Display the medial and lateral menisci and their ligaments on the tibial articular surfaces.

Open the proximal tibiofibular joint.

Ankle joint. Complete the dissection of the following muscles: tibialis anterior, tibialis posterior, peroneus longus, and peroneus brevis. Reflect the muscles toward their digital attachments and observe the anterior aspect of the ankle joint (p. 438).

Remove the remaining parts of the retinacula, together with the tendons of the flexor digitorum longus, flexor hallucis longus, extensor digitorum longus, and extensor hallucis longus muscles, from around the ankle joint.

Dissect the deltoid ligament on the medial side of the joint, from the medial malleolus to the calcaneus and navicular bones (p. 439).

Dissect the anterior talofibular and calcaneofibular ligaments on the lateral side of the joint, from the lateral malleolus to the head of the talus and the calcaneus, respectively (p. 441).

Expose the posterior tibiofibular, transverse, and the posterior talofibular ligaments on the posterior aspects of the distal tibiofibular and ankle joints (p. 440).

Cut the deltoid ligament horizontally, reflect the foot

Table 41

Name	Origin	Insertion	Innervation	Action
Adductor magnus (see Table 35)				
Articularis genu	Anterior surface of lower part of femoral shaft	Upper surface of synovial membrane of knee joint	Femoral nerve	Elevates synovial capsule during extension of leg
Biceps femoris (see Table 36)				
Extensor digitorum longus (see Table 40)				
Extensor hallucis longus (see Table 40)				
Flexor digitorum longus (see Table 38)				
Flexor hallucis longus (see Table 38)				
Gastrocnemius (see Table 37)				
Gracilis (see Table 35)				
Iliopsoas (see Table 28)				
Peroneus brevis (see Table 38)				
Peroneus longus (see Table 38)				
Plantaris (see Table 38)				
Popliteus (see Table 38)				
Quadriceps femoris (see Table 34)				
Sartorius (see Table 34)				
Semimembranosus (see Table 36)				
Semitendinosus (see Table 36)				
Tibialis anterior (see Table 40)				
Tibialis posterior (see Table 38)				

toward the lateral malleolus, and observe the articular surfaces of the ankle joint.

Tarsal and tarsometatarsal joints. Expose the interosseous talocalcaneal ligament in the tarsal sinus on the dorsal aspect of the tarsus (p. 442).

Expose the spring—the plantar calcaneonavicular ligaments on the plantar surface of the tarsus (p. 443).

Cut the articular capsules and ligaments to separate the tarsal bones from one another and the tarsal from the metatarsal bones. Study the articular surfaces.

Metatarsophalangeal and interphalangeal joints. Open all metatarsophalangeal and interphalangeal joints.

NOTE. In the course of this dissecting procedure all muscular attachments should be observed and, in all joints dissected, the articular surfaces and movements studied.

Summary of structures

1. Bones and bony landmarks
   - Acetabulum
   - Anterior inferior iliac spine
   - Calcaneus
   - Femur
   - Femur, head
   - Greater trochanter
   - Hip
   - Lateral malleolus
   - Lesser trochanter
   - Medial malleolus
   - Metatarsals
   - Navicular
   - Patella

Talus, head
Tarsal sinus
Tarsals
Tarsus
Tibia
Tibia, tuberosity

2. Bursae
   Iliopsoas
   Infrapatellar
3. Joints
   Ankle
   Articular capsule
   Articular surface
   Distal tibiofibular
   Hip
   Interphalangeal
   Knee
   Metatarsophalangeal
   Proximal tibiofibular
   Tarsal
   Tarsometatarsal
4. Ligaments
   Anterior cruciate
   Anterior talofibular
   Calcaneofibular
   Deltoid
   Fibular collateral
   Head of femur
   Iliofemoral
   Ischiofemoral
   Long plantar
   Meniscal
   Oblique popliteal
   Patellae
   Posterior cruciate
   Posterior talofibular
   Posterior tibiofibular
   Pubofemoral
   Short plantar
   Spring (plantar calcaneonavicular)
   Talocalcaneal, interosseous
   Tibial collateral
   Transverse, ankle
   Transverse, hip
5. Menisci
   Lateral
   Medial
6. Muscles
   Adductor magnus
   Adductors
   Articularis genu
   Biceps femoris
   Extensor digitorum longus
   Extensor hallucis longus
   Flexor digitorum longus
   Flexor hallucis longus
   Gastrocnemius
   Gracilis
   Iliopsoas
   Peroneus brevis
   Peroneus longus
   Plantaris
   Popliteus
   Quadriceps femoris, heads
   Rectus femoris
   Sartorius
   Semimembranosus
   Semitendinosus
   Tibialis anterior
   Tibialis posterior
7. Retinacula
   Extensor
   Flexor

Alphabetical index

U

V

W

X

Z

Categorical index

Autonomic nervous system

Bones and bony structures

Cartilages and fibrocartilages

Central nervous system and meninges

Digestive system, peritoneum, and spleen

Ear

Glands and excretory ducts

Heart and pericardium

Joints

Larynx

Ligaments

Lymphoid tissue and lymphatic vessels

Muscles, tendons, and bursae

Nerves

Nose and paranasal sinuses

Ocular and lacrimal apparatus

Pharynx

Reproductive system

Respiratory system and pleura

Surface landmarks

Tongue

Topographic regions

Urinary system

Veins